The Kestrel

Widespread across open lands and cities of Europe, Africa and Asia, the common kestrel (*Falco tinnunculus*) is one of the most abundant and studied birds of prey. This book brings together and synthesises the results of research on kestrels for professional ornithologists and scientists that seek to consolidate a vast body of literature. It is also a reference for those readers who may not have the depth of scientific knowledge to navigate new fields of scientific enquiry. It examines many aspects of the species' biology, from the reproductive strategies to the behavioural and demographic adaptations to changes of environmental conditions. It also discusses the roles of physiology and immunology in mediating the adaptability of kestrels to the ongoing environmental changes with a particular focus on contaminants. This volume presents new and exciting avenues of research on the ecology and behaviour of the common kestrel.

David Costantini is Professor of Conservation Physiology at the Muséum National d'Histoire Naturelle in Paris, France. His main research interests have focussed on the physiological mechanisms mediating life-history variation and functions in animal ecology and conservation.

Giacomo Dell'Omo is President of *Ornis italica*, a non-profit Italian association for research and education on environmental issues. His main research interests have focused on the behaviour, ecology and ecotoxicology of birds. He is currently involved in projects relating to many species including kestrels as a favourite.

The Kestrel

Ecology, Behaviour and Conservation of an Open-Land Predator

DAVID COSTANTINI
Muséum National d'Histoire Naturelle, Paris

GIACOMO DELL'OMO
Ornis italica

CAMBRIDGE
UNIVERSITY PRESS

University Printing House, Cambridge CB2 8BS, United Kingdom

One Liberty Plaza, 20th Floor, New York, NY 10006, USA

477 Williamstown Road, Port Melbourne, VIC 3207, Australia

314–321, 3rd Floor, Plot 3, Splendor Forum, Jasola District Centre,
New Delhi – 110025, India

79 Anson Road, #06–04/06, Singapore 079906

Cambridge University Press is part of the University of Cambridge.

It furthers the University's mission by disseminating knowledge in the pursuit of
education, learning, and research at the highest international levels of excellence.

www.cambridge.org
Information on this title: www.cambridge.org/9781108470629
DOI: 10.1017/9781108692120

© David Costantini and Giacomo Dell'Omo 2020

First published 2020

Printed in the United Kingdom by TJ International Ltd, Padstow Cornwall

A catalogue record for this publication is available from the British Library.

Library of Congress Cataloging-in-Publication Data
Names: Costantini, David, author. | Dell'Omo, Giacomo, author.
Title: The kestrel : ecology, behaviour and conservation of an open-land predator / David Costantini,
Muséum National d'Histoire Naturelle, Paris, Giacomo Dell'Omo, Ornis Italia.
Description: Cambridge, United Kingdom ; New York, NY : Cambridge University Press, 2020. |
Includes index.
Identifiers: LCCN 2020007256 (print) | LCCN 2020007257 (ebook) | ISBN 9781108470629
(hardback) | ISBN 9781108692120 (ebook)
Subjects: LCSH: Kestrels.
Classification: LCC QL696.F34 C675 2020 (print) | LCC QL696.F34 (ebook) | DDC 598.9/6–dc23
LC record available at https://lccn.loc.gov/2020007256
LC ebook record available at https://lccn.loc.gov/2020007257

ISBN 978-1-108-47062-9 Hardback

Contents

The plate section can be found between pages 118 and 119.

Preface

The common kestrel is probably the bird of prey most of us are more familiar with. Who can say they have never seen one of these beautiful birds wind-hovering in the countryside or even in the city nowadays? Apart from its beauty, the common kestrel and some of its closely related species have played a central role as study species in behavioural ecology, ecophysiology and environmental toxicology over the last 40 years. Most of this work has been carried out just after the publication of the seminal monography *The kestrel* by Andrew Village in 1990, which has been a reference for all those, like us, that have been fascinated by this species. A few other monographic books on the kestrel were published in the last decades and are worth mentioning, such as *Season with the kestrel* by Gordon Riddle in 1991, *The kestrel* by Michael Shrubb in 1993 or *Pustułka* by Śliwa and Rejt in 2006. These books are, however, more divulgative and Village's book remains the main scientific text of reference.

We have followed our passion for kestrels for some 20 years by setting up a study population breeding in nest boxes attached to pylons of utility lines in central Italy. Together with our colleague and friend Stefania Casagrande, we have monitored the nests and carried out our first observations and experimental work. Since then, our passion has not faded; we continue to maintain our high scientific interest in kestrels. We are aware that other groups in Europe have historical dedication and competences for this species; however, no one has produced an updated synthesis of the research on the kestrel, and Village's text has remained the main comprehensive scientific contribution to the species. Thus, we think that the time is ripe for a new book that synthesises the substantial literature published so far on the common kestrel and integrates it with some novel data. Although this book is focused on the common kestrel, we often refer to work on other kestrel species or even different species in order to give a comparative framework to address specific questions. The book has not the presumption to be exhaustive in terms of literature coverage and discussion of results. Hence, we advise readers to refer to the primary literature quoted in this book when needed. We have learned much about the behaviour and the ecology of the kestrel, but much of this knowledge comes from studies limited to a few areas that cover only a small part of the whole range of the species. Recognising such limitations will be instructive and will offer new stimuli for replication of experiments under different environmental contexts and experimental opportunities to explore novel questions.

1 Systematics and Evolution of Kestrels

1.1 Chapter Summary

The family Falconidae constitutes a group of small to medium-sized diurnal raptors whose monophyly is strongly supported. Kestrels are included in the subfamily Falconinae. There are at least 13 species that belong to the kestrel group, but recent genetic studies suggest that the number of kestrel species might be larger, possibly 16. The paleontological and molecular evidence is congruent in suggesting an evolutionary radiation of kestrels from the Late Miocene (4.0–9.8 million years ago) through the Early Pleistocene. However, the geographic area where kestrels originated and dispersed from is unclear.

1.2 Diversification of Falcons

The Falconidae is a monophyletic family of diurnal birds of prey that occupy a wide variety of ecological niches and geographic regions (White et al., 1994). Three subfamilies are currently recognised and their validity is supported by both molecular and morphological data (Griffiths, 1999; Griffiths et al., 2004; Fuchs et al., 2012, 2015): (i) Falconinae (falcons, falconets and kestrels), (ii) Herpetotherinae (forest falcons *Micrastur* sp. and laughing falcon *Herpetotheres cachinnans*) and (iii) Polyborinae (caracaras) (Figure 1.1). Dickinson (2003) has recognised 11 genera and 64 species of Falconidae, but figures can vary slightly across authors.

Both the Herpetotherinae and the Polyborinae occur only in the New World, while the Falconinae (the subfamily to which kestrels belong) are widespread across both the New and Old World with 46 species, 40 of which belong to the genus *Falco* (Fuchs et al., 2015). Molecular genetic estimates of diversification within Falconidae, based on fossil calibration using two *Falco* ancestors (*Pedohierax* and *Thegornis*) and on the analyses of DNA sequences from eight loci, indicated that the diversification started between 22.3 (95% confidence interval: 19.6–25.6) and 34.2 (95% confidence interval: 26.2–43.2) million years ago for the Falconidae; between 12.6 (95% confidence interval: 10.6–14.7) and 19.3 (95% confidence interval: 14.4–24.6) million years ago for the

Figure 1.1 The family *Falconidae* includes three subfamilies: (a) Falconinae (*Falco tinnunculus*, photograph by David Costantini); (b) Herpetotherinae (*Herpetotheres cachinnans*, photograph by Andreas Trepte, via Wikimedia Commons); (c) Polyborinae (*Caracara cheriway*, photography by www.naturespicsonline.com, via Wikimedia Commons).

Falconinae; and between 5.0 (95% confidence interval: 4.0–6.1) and 7.7 (95% confidence interval: 5.6–9.8) million years ago for the genus *Falco* (Figure 1.2; Fuchs et al., 2015).

Cenizo et al. (2016) proposed that the Lower Eocene *Antarctoboenus carlinii*, a fossil species found on Seymour island (west Antarctica), could represent the most ancient falconiform described so far (Figure 1.3); this discovery would give support to a Neotropical or Austral origin of Falconidae (Ericson et al., 2006; Ericson, 2012; Fuchs et al., 2015).

The diversification of the genus *Falco* started during a period characterised by increasing aridity and the spread of open savannahs (Cerling et al., 1997), which might have favoured the diversification of these open-land birds of prey (Cade & Digby, 1982). Molecular estimates of the diversification of species belonging to the genus *Falco* provided by Fuchs et al. (2015) are in agreement with the paleontological evidence, as fossils of several *Falco* paleospecies date from the Late Miocene to the Early Pliocene period (e.g. Umanskajaa, 1981; Becker, 1987; Boev, 1999, 2011a, 2011b, 2011c; Li et al., 2014). However, most *Falco* paleospecies are known only from fragmentary remains, which make phylogenetic inferences problematic. In conclusion, the recent molecular and paleontological data have supported the statement made by Cade and Digby (1982):

The Late Miocene or Early Pliocene would seem to have been about the right time, just when things were starting to go well for another group of open-country inhabitants, the early hominids. It is amusing and somehow prophetic to think that falcons and men both derive from the same evolutionary stimulus – the creation of open grasslands and savannahs with new and unexploited opportunities for both winged and bipedal hunters … Thus it appears that the association between men and falcons is deep rooted indeed. What did 'Lucy' and her kin (*Australopithecus afarensis*) experience when they looked up into the azure sky over the Afar Plains and saw hunting falcons?

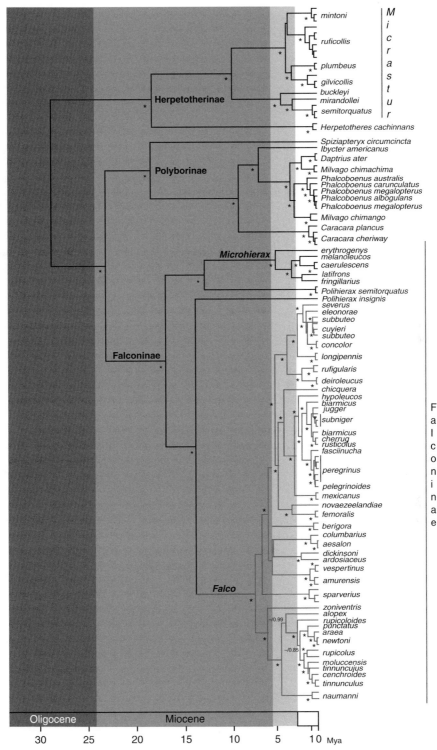

Figure 1.2 Chronogram calculated using eight genetic markers and two fossil taxa (*Pedohierax* and *Thegornis*) as calibration points. Asterisks indicate posterior probabilities and maximum likelihood bootstrap support values higher than 0.95 and 70%, respectively. The different shades of grey indicate different geological epochs. The different branch colours indicate the two parts of the tree with different diversification rates. Reprinted from Fuchs et al. (2015) with permission from Elsevier.

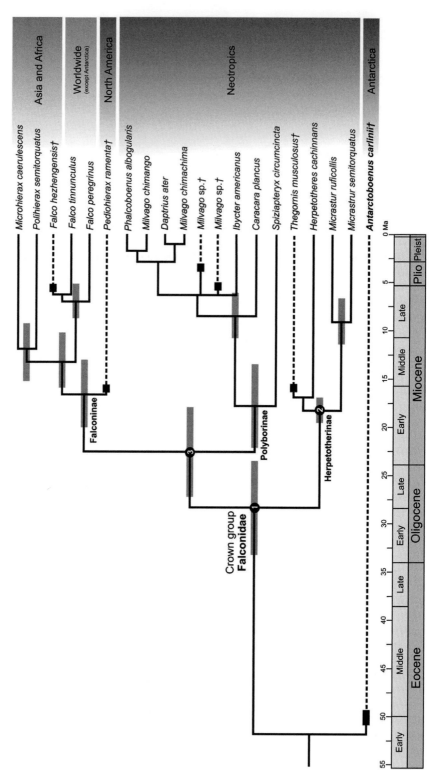

Figure 1.3 Geographical range, temporal distribution, and phylogenetic affinities of extant and fossil falconid birds. The temporal distribution of fossil taxa (dagger) is indicated by black squares. Grey bars indicate divergence times estimates for the primary lineages within the extant Falconids according to Fuchs et al. (2015). Reprinted from Cenizo et al. (2016) with permission from Springer Nature.

1.3   Systematics and Diversification of Kestrels

1.3.1   Morphological and Behavioural Evidences

Kestrels belong to the subfamily Falconinae. It is traditionally recognised that at least 13 species belong to the kestrel group, but recent molecular analyses have suggested that this number of species might be larger (Table 1.1).

Boyce and White (1987) suggested that there are 14 or 15 species of *Falco* that might be considered as kestrels. The classic kestrel group includes a single New World species (the American kestrel, *F. sparverius*) and 12 Old World species. The majority of kestrels (10 of 13 species) have been classified within the subgenus *Tinnunculus* (Brown & Amadon, 1968; Cade & Digby, 1982). This subgenus includes those species characterised by a classic kestrel form and brown-rufous plumage colouration. The three species characterised by grey plumage colouration – the grey kestrel (*F. ardosiaceus*), Madagascar banded kestrel (*F. zoniventris*) and Dickinson's kestrel (*F. dickinsoni*) – are all from Africa and have been placed within the subgenus *Dissodectes* (Snow, 1978). This distinction between *Tinnunculus* and *Dissodectes* kestrels has also been supported by the electrophoretic patterns of feather proteins (Olsen et al., 1989).

Table 1.1 List of species that belong or have been suggested to belong to the kestrel group. 1 = Fuchs et al. (2015) have proposed that the subspecies *rupicolus* may warrant species status based on molecular genetic analyses. 2 = Brown and Amadon (1968) and Fuchs et al. (2015) suggested that *F. amurensis* and *F. vespertinus* might be part of the kestrel group.

Common name	Scientific name	Authority
Common kestrel	*Falco tinnunculus*	Linnaeus, 1758
Lesser kestrel	*Falco naumanni*	Fleischer, 1818
Fox kestrel	*Falco alopex*	Heuglin, 1861
Seychelles kestrel	*Falco araea*	Oberholser, 1917
Grey kestrel	*Falco ardosiaceus*	Vieillot, 1823
Dickinson's kestrel	*Falco dickinsoni*	Sclater, 1864
Moluccan or spotted kestrel	*Falco moluccensis*	Bonaparte, 1850
Madagascar kestrel	*Falco newtoni*	Gurney, 1863
Mauritius kestrel	*Falco punctatus*	Temminck, 1821
Greater kestrel	*Falco rupicoloides*	Smith, 1829
Madagascar banded kestrel	*Falco zoniventris*	Peters, 1854
Australian nankeen kestrel	*Falco cenchroides*	Vigors & Horsfield, 1827
American kestrel	*Falco sparverius*	Linnaeus, 1758
Rock kestrel	*Falco rupicolus*[1]	Daudin, 1800
Red-footed falcon	*Falco vespertinus*[2]	Linnaeus, 1766
Amur falcon	*Falco amurensis*[2]	Radde, 1863

There are no clear diagnostic features that define what a kestrel is and clarify the evolutionary relationships among kestrel species. For example, while many kestrel species perform hovering for hunting small mammals, this hunting technique appears to be very uncommon in the grey kestrel, Madagascar kestrel (*F. newtoni*) or fox kestrel (*F. alopex*) (Gaymer, 1967; Boyce & White, 1987; Londei, 2002). However, we know very little about the behaviour and the ecology of kestrels in Africa and Asia, which limits the reliability of any conclusions. Sexual dimorphism in plumage is also not common across all kestrel species. For example, sexual dimorphism is apparent in the common kestrel (*F. tinnunculus*), the lesser kestrel (*F. naumanni*) and the Australian nankeen kestrel (*F. cenchroides*), while it is barely evident in other species. As for the plumage colouration in nestlings, while they usually resemble adult females, male and female American kestrel nestlings differ in plumage colouration (Village, 1990).

Phylogenetic analyses carried out using data sets of morphological (e.g. plumage colouration, wing size) and behavioural (e.g. hovering) traits led Boyce and White (1987) to conclude that the fox kestrel might be the ancestor species of the current red and grey kestrel species, mainly because it does not hover and has negligible age or colour dimorphism between sexes. This conclusion was also based on the assumption that both the lack of hovering and the negligible colour dimorphism are primitive characters, i.e. closer to the most ancestral kestrel species. This might be plausible if the ancestor species were a raptor living in forests, where hovering would not have been an efficient hunting strategy. However, the fox kestrel can hover (Londei, 2002) and the phylogenetic value of hovering is unclear. Boyce and White (1987) also concluded that the American kestrel might be the most recently evolved species of kestrel because (i) it has strong adult sexual dimorphism; (ii) male and female nestlings differ in colouration, with each sex resembling their respective adult sex rather than the female; and (iii) the species occupies an extensive geographic area with 17 subspecies (14 in Boyce & White, 1987) that have not diverged enough to be considered distinct species. Finally, the analyses further suggested that the lesser kestrel might share a direct ancestor species with the Amur falcon (*F. amurensis*) and the red-footed falcon (*F. vespertinus*).

Reconstruction of phylogenetic trees based on morphological characters may be questioned because two species might be similar to each other because of convergent evolution; this occurs when species share a trait that is different from the trait inferred to have been present in their common ancestor. Thus, the same trait has evolved independently in the two species in order to perform a similar function. Other problems in using morphological traits may arise with insular species. For example, the negligible sexual dimorphism of the Mauritius kestrel (*F. punctatus*) might (i) represent the ancestral condition, (ii) indicate the loss of dimorphism due to the insular isolation or (iii) suggest that barely dimorphic individuals were those that colonised the islands. Also, compared to other kestrel species, the Mauritius kestrel has more rounded wings (like those of forest-dwelling raptors), probably because of its adaptations to hunt in forested habitats. Thus, molecular data (e.g. DNA sequences of marker genes) are needed to elucidate evolutionary relationships because they are less biased by convergent evolution than are morphological traits.

1.3.2   Molecular Evidence

Finer subdivisions within the *Tinnunculus* kestrels have been contentious for years. Groombridge et al. (2002) constructed a molecular phylogeny of *Tinnunculus* kestrels using the mitochondrial cytochrome *b* DNA sequence. All analytical approaches used by Groombridge et al. (2002) produced phylogenetic trees of broadly similar topology, but with inconsistent positions for the Mauritius kestrel and the greater kestrel (*F. rupicoloides*). The molecular analyses provided strong support for a common ancestor shared between the Madagascar kestrel and the Seychelles kestrel (*F. araea*). The molecular data indicated that the kestrels that colonised the Seychelles likely came from Madagascar between 0.3 and 1.0 million years ago, probably favoured by the lower sea level due to glaciation (Groombridge et al., 2002). Given this scenario, it has been suggested that the Aldabran kestrel (*F. newtoni aldabranus*), a subspecies of the Madagascar kestrel (Benson, 1967; Benson & Penny, 1971), might be a possible relict of such dispersal (Groombridge et al., 2002). It has also been hypothesised that the colonisation of Mauritius occurred via Madagascar less than 3 million years ago (Groombridge et al., 2002). The combination of molecular and geological data led to the hypothesis that the colonisation of Mauritius by kestrels occurred about 1.9–2.6 million years ago, a period characterised by an interruption in the volcanic activity on Mauritius (Groombridge et al., 2002). As compared to the Madagascar kestrel, both the Seychelles kestrel and the Mauritius kestrel show morphological adaptations for forest-dwelling. Such adaptations might have evolved on the islands after the dispersion from Madagascar ended (Groombridge et al., 2002). Alternatively, it might be that individual Madagascar kestrels with a more forest-kestrel form might have been the pioneers of such colonisation.

Molecular data produced by Groombridge et al. (2002) suggest a close affinity between the common kestrel and the Australian nankeen kestrel, supporting a recent divergence, probably due to Pleistocene glacial events that pushed the common kestrel stock southwards from Asia, as previously hypothesised by Boyce and White (1987) on the basis of morphological traits. Further work based on the mitochondrial cytochrome *b* gene clustered the red-footed falcon, the Amur falcon, Dickinson's kestrel and the American kestrel separately from the kestrel group (Wink & Sauer-Gürth, 2004).

More recent molecular analyses made by Fuchs et al. (2015) did not provide support for the traditional hypothesis proposing (i) the American kestrel as closely related to the Old World kestrels and (ii) the Madagascar banded kestrel as closely related to the grey kestrel and Dickinson's kestrel. Rather, Fuchs et al. (2015) hypothesised (i) the existence of a group including the Old World kestrels together with the Madagascar banded kestrel and (ii) that the American kestrel would be more closely related to other falcons (e.g. red-footed falcon) than to the Old World kestrels, as previously suggested by DNA/DNA hybridisation studies (Sibley & Ahlquist, 1990) and by analyses of mitochondrial cytochrome *b* DNA sequence (Groombridge et al., 2002). These results might indicate that either the kestrel group should include more than the 13 species classically recognised or that the position of the American kestrel within the kestrel group might need to be reconsidered. Analyses of molecular genetic markers made by Fuchs et al. (2015) also suggested that the Amur falcon and the red-footed falcon might be part of the kestrel group, as previously proposed by Brown and Amadon (1968). Finally, Fuchs et al. (2015) found that the fox kestrel might not be the sister species of the greater kestrel as traditionally thought. Rather, both the fox kestrel and the greater kestrel would form a paraphyletic group (i.e. a group which does not include all descendants of the same ancestor), with the greater kestrel being sister to all other kestrels included in the *F. tinnunculus* group.

1.3.3   Paleontological Evidence

There is scarce information about the fossil history of raptors belonging to the genus *Falco*, particularly because of poor preservation of bone remains. Two close relative species of the common kestrel are the Late Pliocene *F. bakalovi*, whose remains were found near the town of Varshets in northwest Bulgaria (Boev, 1999, 2011a, 2011b), and the Late Miocene *F. bulgaricus*, whose remains were found near the town of Hadzhidimovo in southwest Bulgaria (Boev, 2011c). As with other *Falco* fossils, both *F. bakalovi* and *F. bulgaricus* have been described on the basis of a limited number of isolated bones, which makes their description and systematic classification problematic. One of the best-preserved specimens of a fossil falconid so far recovered has been classified as *F. hezhengensis* (Figure 1.4) by Li et al. (2014).

This falconid species was found in the Late Miocene deposits of Linxia Basin in north-western China. The analysis of 66 osteological characters led the authors to place this new *Falco* paleospecies as the sister group of the common kestrel and the greater kestrel (Li et al., 2014). This new paleospecies has also provided paleontological evidence in favour of an earlier divergence of kestrels from the peregrine falcon (*F. peregrinus*) in the Late Miocene. Although the paleontological evidence is congruent with the molecular data in suggesting a radiation of kestrels from the Late Miocene through the Early Pleistocene, the timing and location of the radiation are still ambiguous because of the paucity of well-preserved fossils over a wide geographic area. It has been suggested that the presence of the majority of kestrel taxa on the African continent could be evidence for an African origin of the kestrel group. However, the absence of a Pre-Pleistocene kestrel fossil record from Africa has meant that assumptions of kestrel divergence within and from the African continent have not been easy to confirm. Li et al. (2014) suggested that *F. hezhengensis* might represent an early dispersal event after the origin of kestrels occurred or, alternatively, it might be interpreted as evidence in favour of an Eurasian origin of kestrels.

1.4   Diversification and Geographic Distribution of Kestrels

The kestrel group is truly cosmopolitan, occupying wide geographical areas and different environments. Most kestrels are open-land predators, and avoid deserts,

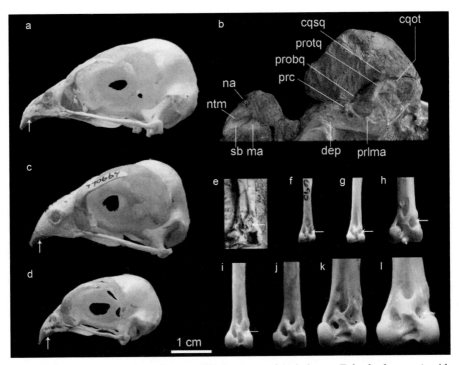

Figure 1.4 Comparison of the skull and distal tibiotarsus of the holotype *Falco hezhengensis* with other Falconidae species: (a) *F. tinnuculus* (USNM 610371); (b, e) *F. hezhengensis* (IVPP V14586); (c) *Polihierax insignis* (USNM 490664); (d, f) *Microhierax erythrogenys* (USNM 613695); (g) *P. semitorquatus* (USNM 621024); (h) *F. rupicoloides* (USNM 430626); (i) *Spiziapteryx circumcincta* (USNM 319445); (j) *Micrastur ruficollis* (USNM 621387); (k) *Herpetotheres cachinnans* (USNM 346714); and (l) *Caracara plancus* (USNM 614583). The vertical arrows indicate the tomial notch on the premaxilla that is present in Falconinae; the horizontal arrows indicate a third opening into the extensor groove located medial to the supratendinal bridge that is present in Falconinae and Polyborinae but not in Herpetotherinae. Anatomical abbreviations: cqot, cotyla quadrati otici; cqsq, cotyla quadrati squamosa; dep, depression; ma, mandible; na, naris; ntm, notched tomial margin; prc, processus coronoideus; prlma, processus lateralis mandibulae; probq, processus orbitalis quadrati; protq, processus oticus quadrati; sb, septal bar. Reprinted from Li et al. (2014) with permission from the American Ornithological Society.

dense forests and the Arctic–Antarctic poles. Some kestrel species show large geographic variation in both morphology and genetics, which led to the recognition of several subspecies. The higher number of species and subspecies near the equator might be due to greater speciation favoured by a more sedentary style of local birds as compared to kestrels at higher latitudes, whose migratory habits limit isolation among populations. There are, however, a number of factors that have likely contributed to the diversification of kestrels, such as physical barriers and geographic isolation (Village, 1990). We currently lack studies that elucidate the roles that different barriers had in driving the diversification of kestrels.

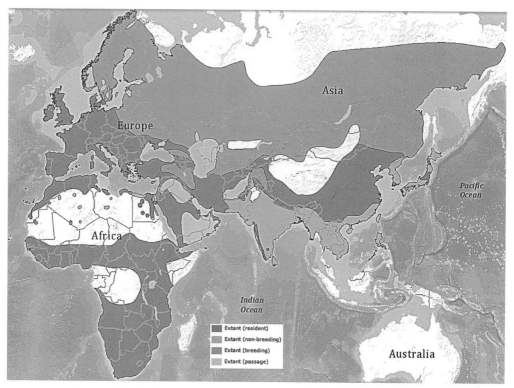

Figure 1.5 Distribution map of the common kestrel (*Falco tinnunculus*). From BirdLife International and Handbook of the Birds of the World (2017). Reproduced with permission from BirdLife International. A black and white version of this figure will appear in some formats. For the colour version, refer to the plate section.

The common (also known as Eurasian) kestrel is widespread across Europe, Africa and Asia (Figure 1.5), but sporadic observations of free-ranging kestrels have been recorded in the New World (Campbell, 1985; Pranty et al., 2004).

It is classically recognised that there are 11 subspecies of common kestrel (Village, 1990). However, recent molecular analyses led Fuchs et al. (2015) to propose that the subspecies *rupicolus* (Table 1.1) may warrant species status (IOC World Bird List: Gill & Donsker, 2018). The species or subspecies *rupicolus* (illustrated in Figure 1.6b) occurs from north-western Angola and southern Democratic Republic of Congo to southern Tanzania and South Africa. The subspecies belonging to the *tinnunculus* group differ in colouration, body size and distribution (Figures 1.6a,b). The nominal subspecies *tinnunculus* breeds from Europe and north Africa east to Siberia, Bhutan and western China. It is partially migratory, wintering in the regions from south to central Africa, India and south-eastern Asia (Figure 1.6a; Village, 1990; BirdLife International and Handbook of the Birds of the World, 2017). Four subspecies are endemic to the Atlantic islands of Macaronesia. The subspecies *neglectus* inhabits the north Cape Verde islands and the subspecies *alexandri* inhabits the south-east Cape

Figure 1.6 Comparison of subspecies of *Falco tinnunculus* and of other kestrel species. Panel a from left to right: *F. t. tinnunculus* (male, C.G. 1967 N. 1709, MNHN); *F. t. tinnunculus* (female, C.G. 1912 N. 560, MNHN); *F. t. interstinctus* (female, C.G. 2003 N. 159, MNHN); *F. t. rufescenes* (female, C.G. 1977 N. 347, MNHN); *F. t. alexandri* (female, C.G. 1966 N. 904, MNHN); *F. t. neglectus* (female, C.G. 1967 N. 1756, MNHN). Panel b from left to right: *F. t. dacotiae* (female, C.G. 1965 N. 1492, MNHN); *F. t. canariensis* (female, C.G. 1911 N. 882, MNHN); *F. t. canariensis* (male, C.G. 1965 N. 1484, MNHN); *F. moluccensis* (female, C.G. 1882 N. 152, MNHN); *F. rupicolus* (male, C.G. 2018 N. 503, MNHN); *F. r. rupicoloides* (female, C.G. 2003 N. 165, MNHN). Specimens are from the collection of the Muséum National d'Histoire Naturelle (MNHN; Paris, France). The MNHN gives access to the collections in the framework of the RECOLNAT national Research Infrastructure. Photographs by David Costantini. A black and white version of this figure will appear in some formats. For the colour version, refer to the plate section.

Verde islands (Bourne, 1955). The subspecies *alexandri* is similar in size to both subspecies from the Canary Islands, but differs from *neglectus* by being larger, darker and more rufous on the back (Figure 1.6a; Alexander, 1898; Bourne, 1955; Vaurie, 1961). The subspecies *canariensis* inhabits the Madeira and western Canary Islands and the subspecies *dacotiae* inhabits the eastern Canary Islands. These two subspecies are similar in size, but *dacotiae* is paler, redder and less marked than *canariensis*

(Figure 1.6b; Koenig, 1890; Hartert, 1912–21; Vaurie, 1961). The two Canary subspecies also show a significant genetic differentiation (Groombridge et al., 2002; Alcaide et al., 2009).

The subspecies *rupicolaeformis* occurs in north-east Africa and Arabia. Grant and Mackworth-Praed (1934) observed that (i) the male *rupicolaeformis* was more rufous, less pinkish on upper parts and rather more heavily spotted than the male *tinnunculus* and (ii) the female *rupicolaeformis* was paler than the female *tinnunculus* and warmer in colour, tinged with richer rufous, upper tail coverts and base of tail washed with dove grey. Vaurie (1961) found that *rupicolaeformis* tended to be darker, smaller and more spotted or barred than the subspecies *tinnunculus*. He also noticed that the back of male *rupicolaeformis* was more vinaceous and less bright than that of male *tinnunculus*. The subspecies *interstinctus* occurs in China and Japan, and winters in India, Malaysia and the Philippines (Figure 1.6a). Vaurie (1961) observed that (i) *interstinctus* was similar in size to *tinnunculus* and larger than *rupicolaeformis* and (ii) was darker and more spotted, barred and streaked than *tinnunculus*. These similarities and differences between *tinnunculus* and *interstinctus* are also evident in Figure 1.6a. There is also a significant genetic differentiation between the two subspecies, which might have been favoured by the glacial events that occurred during the Quaternary period (Zhang et al., 2008). The subspecies *objurgatus* occurs in south India (western and eastern Ghats) and Sri Lanka. The subspecies *archeri* occurs in Somalia, coastal Kenya and Socotra; it is similar to the subspecies *tinnunculus* but smaller. Finally, the subspecies *rufescens* is widespread from west Africa to Ethiopia, Tanzania and Angola (Figure 1.6a; Village, 1990; BirdLife International and Handbook of the Birds of the World, 2017). Grant and Mackworth-Praed (1934) observed that it is a dark subspecies; however, this is not very evident from the specimen illustrated in Figure 1.6a.

The Mauritius, Madagascar, Madagascar banded, Seychelles and Australian nankeen kestrels are all regarded as distinct species endemic to their respective islands. Two subspecies can be recognised for the Madagascar kestrel (*newtoni* in Madagascar and *aldabranus* in the Aldabra islands) and for the Australian nankeen kestrel (*cenchroides* in Australia and *baru* in New Guinea). The fox kestrel occurs in the northern regions of sub-Saharan Africa (Figure 1.7a). The greater kestrel is restricted to southern and eastern parts of Africa (Figure 1.7b), where it occurs with three subspecies: *rupicoloides* in South Africa (illustrated in Figure 1.6b), *arthuri* in east Africa and *fieldi* in Somalia. The grey kestrel occurs in north Africa (Figure 1.7c) and Dickinson's kestrel occurs in south-east Africa (Figure 1.7d).

The American kestrel is widespread across the whole American continent with 17 subspecies: *sparverius* from Alaska to Newfoundland, south to west Mexico; *peninsularis* in south Baja California, Sonora and Sinaloa; *tropicalis* in south Mexico to north Honduras; *nicaraguensis* in the savannas of Honduras and Nicaragua; *paulus* from the southern coasts of the USA to Florida; *dominicensis* in Hispaniola and Jamaica; *caribaearum* in the West Indies (Puerto Rico to Grenada); *brevipennis* in Aruba, Curaçao and Bonaire; *isabellinus* from Venezuela to north Brazil; *aequatorialis* in north Ecuador; *peruvianus* in south Ecuador, Peru and north Chile; *fernandensis* in the Juan Fernández Islands off Chile; *cinnamominus* in south-east Peru, Chile and

Figure 1.7 Distribution maps of the (a) fox kestrel (*F. alopex*), (b) greater kestrel (*F. rupicoloides*), (c) grey kestrel (*F. ardosiaceus*) and (d) Dickinson's kestrel (*F. dickinsoni*). From BirdLife International and Handbook of the Birds of the World (2017). Reproduced with permission from BirdLife International.

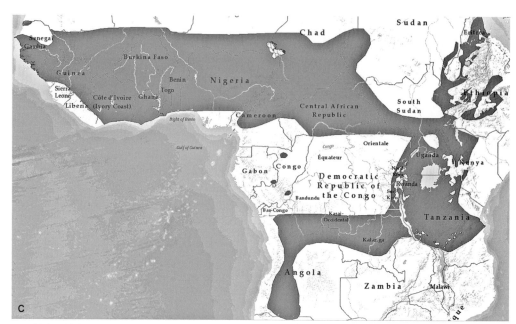

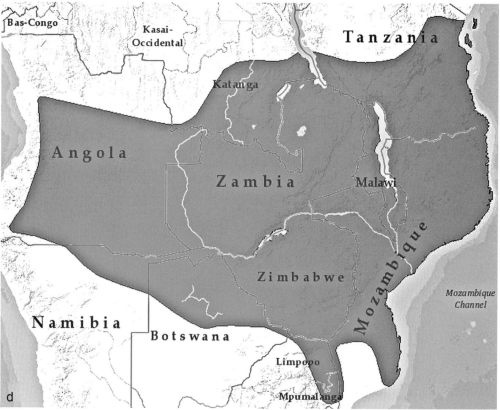

Figure 1.7 (cont.)

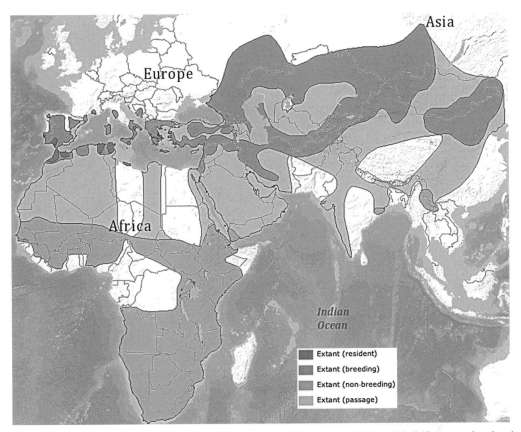

Figure 1.8 Distribution map of the lesser kestrel (*F. naumanni*). From BirdLife International and Handbook of the Birds of the World (2017). Reproduced with permission from BirdLife International. A black and white version of this figure will appear in some formats. For the colour version, refer to the plate section.

Argentina to Tierra del Fuego; *cearae* in the Tablelands of north-east Brazil to east Bolivia; *sparverioides* in Bahamas, Cuba and the Isle of Pines; *ochraceus* on the mountains of east Colombia and north-west Venezuela; *caucae* on the mountains of west Colombia.

The lesser kestrel has extensive ranges across Europe, Africa and Asia (Figure 1.8). The Moluccan kestrel (*F. moluccensis*; illustrated in Figure 1.6b) is endemic to the East Indies (Figure 1.9), where it occurs with two subspecies: *moluccensis* in north and south Moluccas; *microbalius* from Java to Lesser Sunda Islands, Sulawesi and Tanimbar Islands.

Although not traditionally recognised as kestrel species, recent molecular evidence suggested that the red-footed falcon and the Amur falcon are closely related to kestrels (Fuchs et al., 2015). The red-footed falcon occurs in central Eurasia and winters in sub-Saharan Africa, while the Amur falcon occurs in the steppes of north-east Asia and winters from Malawi to South Africa.

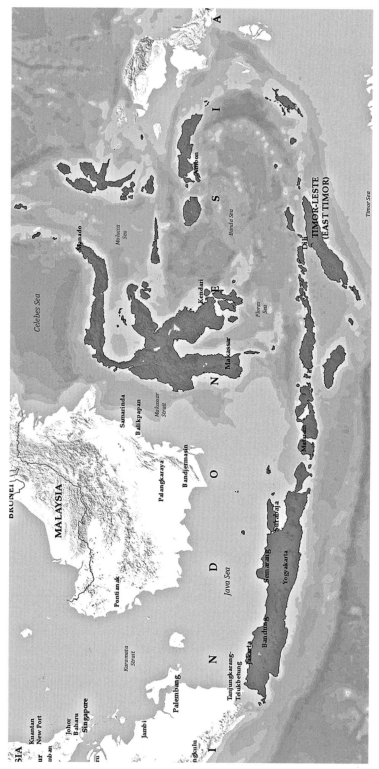

Figure 1.9 Distribution map of the Moluccan kestrel (*F. moluccensis*). From BirdLife International and Handbook of the Birds of the World (2017). Reproduced with permission from BirdLife International.

1.5   Conclusions

Kestrels are a cosmopolitan group, including 13–16 species, depending on the author and the traits used (e.g. morphological, genetic) for reconstructing evolutionary relationships among species. The phylogenetic position of the American kestrel within kestrels is still uncertain, and there is also no agreement on whether the red-footed falcon and the Amur falcon belong to the kestrel group.

The paleontological and molecular evidence is congruent in suggesting an evolutionary radiation of kestrels from the Late Miocene through the Early Pleistocene. However, the absence of a Pre-Pleistocene kestrel fossil record from Africa has not enabled fully clarification of the kestrel divergence within and from the African continent. Finally, more studies of molecular genetics at subspecies level would be needed in order to better clarify the evolutionary history of kestrel species and to define more targeted guidelines for their conservation. For example, 17 subspecies of American kestrel are currently recognised, but limited information is available about their differentiation in terms of behaviour, ecology and genetics (Figueroa & Corales, 2002; Miller et al., 2012; deMent et al., 2014). Such paucity of data makes the geographic boundaries among the subspecies sometimes difficult to identify, as well as the systematic validity of all subspecies difficult to ascertain.

2 Feeding Ecology

2.1 Chapter Summary

The common kestrel is a generalist predator. However, it also shows significant within-species variation in food habits, such as local specialisations on given prey (e.g. voles in northern Europe or lizards in southern Europe) or even individual food preferences. This chapter illustrates the factors that affect the diet composition of kestrels, their foraging strategies and the processes of food competition, including kleptoparasitism. It also explores last-generation techniques, like stable isotope analyses and acceler-ometer–GPS loggers, that would enable us to overcome the limits of the classical methods used to study the feeding ecology of kestrels.

2.2 Introduction

Feeding ecology provides a comprehensive framework for the many factors that shape the patterns of food search and acquisition to its digestion. Food intake is strictly connected with many aspects of an animal's life, such as neonatal mass, the rate at which an individual grows, adult mass gain, body energy reserves, physiological and immunological status, reproductive activity, survival probabilities, foraging behaviour or development of secondary sexual traits (e.g. Byrne et al., 2008; Lee et al., 2008; Eeva et al., 2009; Raubenheimer et al., 2009; Sorensen et al., 2009; Wagner et al., 2013). Our understanding of the mechanisms driving the evolution of feeding ecology and its connections to variation in either life histories or physiology is still limited. Many features characterise the feeding ecology of a species, from the search, selection and ingestion of food to its digestion and absorption of nutrients. This implies that a number of factors play a role in shaping the feeding behaviour.

The common kestrel is an open-land predator (Figure 2.1), generally considered as a food generalist and an opportunistic forager, catching what is locally available. However, recent research showed that there may be significant individual variation in feeding behaviour (Section 2.4.5). This is particularly true for those populations with a large trophic niche. Kestrels can actually feed on a large variety of prey, sometimes even including large birds such as the woodpigeon *Columba palumbus* (Village, 1990) and the magpie *Pica pica* (Costantini et al., 2005) or mammals such as the weasel *Mustela nivalis* (Masman, 1986, p. 32).

Figure 2.1 Open landscapes dominated by grasslands are optimal habitats for kestrels. Cities with monuments and green areas have become novel breeding habitats for kestrels, but their suitability to sustain viable populations of kestrels is uncertain as yet. Photographs by David Costantini.

There has been substantial interest in characterising the diet composition of many European populations of kestrel, while comparatively less is known about the diet of kestrels in other geographic regions. In other words, many conclusions about the feeding ecology of kestrels were derived from data collected from populations of northern Europe. These conclusions appear to hardly apply to populations from lower latitudes for a number of reasons, such as larger trophic niches, occurrence of individual differences in feeding ecology or more stable populations of prey that occur at lower latitudes. Although not yet fully corroborated, kestrels very probably originated in Africa, where kestrel species show large variation in food habits and have a diet composition rich in reptiles and insects (Section 2.4.1). Thus, either kestrels likely became vole specialists in central and northern Europe secondarily after their dispersion northwards or individuals with a propensity to feed on voles might have been those that colonised the northern regions of Europe.

2.3   Analysis of Diet Composition

2.3.1   Classic Methods for Describing the Diet of Kestrels

Scientists have used various methods to describe the diet composition of kestrels. Direct observation of hunting kestrels is certainly the most classical but time-consuming approach. Analysis of undigested remains in pellets regurgitated by the birds or of fresh prey remains makes dietary description much easier. Kestrels produce pellets of undigested parts of their prey that are easy to find under roosts or in nests. They are produced about 24 hours after the meal (Duke et al., 1976, 1996). The analysis of pellets consists in the taxonomic identification of prey and the counting of prey remains (e.g. skulls, left or right jaws for rodents; beaks, left or right humerus for birds; parts of insects) in order to produce a numerical figure of the prey and, eventually, to estimate the biomass eaten. Analyses of pellets and prey remains can also enable scientists to identify the taxonomic order and family or even the species of each prey and its attributes (e.g. sex, age, body size). However, the accuracy of conventional pellet analysis for assessing the diet composition has limitations. It underestimates the actual prey composition and biomass because small or soft-bodied prey are completely digested and may not leave any remains (Yalden & Yalden, 1985; Village, 1990). Also, kestrels leave remains of prey that they do not swallow at the feeding places, for example the fur of small mammals, the feathers of birds, the stomachs of lizards or the heads of passerines and lizards (Masman et al., 1986; Steen et al., 2010; personal observations; Figure 2.2). During the reproductive season, this behaviour (known as prey preparation) is commonly used to prepare prey for the offspring to facilitate their swallowing (Steen et al., 2010). Kestrels also accumulate fresh prey in the nest (Figure 2.3) or in caches (Section 2.7.2). However, comparison of diet composition among studies is not straightforward to achieve, because study methods and their associated drawbacks vary. Also, data on prey abundance and profitability (energy intake per unit time) are not always available, which makes it difficult to provide estimates of diet composition standardised for the available prey. The profitability of a prey may also be influenced by its body size, which would determine the cost of its transportation from the capture site to the nest or to the perch. The bias introduced by the *load-size effect* means that the functional response of a predator (prey capture rate as a function of prey density) or prey selection based on analyses of food remains left in the nests might be problematic to assess (Sonerud, 1992). Other factors are also involved in determining whether the prey will be transported to the nest, which would make the

Figure 2.2 A male kestrel decapitating an Italian three-toed skink (*Chalcides chalcides*). Photograph by Gianluca Damiani. A black and white version of this figure will appear in some formats. For the colour version, refer to the plate section.

load-size effect more or less important. For example, partial consumption of large prey at the capture site may reduce the load-size (Korpimäki et al., 1994). Also, males may face conflicting needs when they provide food to the brooding female and the offspring, which would affect their foraging decisions (Korpimäki et al., 1994). In central Italy, large western green lizards (*Lacerta bilineata*), particularly males, are transported entire to the nest (Figure 2.4) and represent an important component of diet for many breeding kestrels (Costantini et al., 2005, 2007c).

2.3.2   Stable Isotope Ratios to Estimate the Trophic Niche

A complimentary but unexplored method to assess the trophic niche and diet of kestrels is analysis of stable isotope ratios in biological matrices, such as blood or faeces. In nature, both carbon and nitrogen occur as two stable isotopes: ^{12}C and ^{13}C; ^{14}N and ^{15}N. Ratios of the two isotopes as compared with reference standards are noted as $\delta^{13}C$ ($^{13}C/^{12}C$) for carbon and $\delta^{15}N$ ($^{15}N/^{14}N$) for nitrogen, respectively (DeNiro &

Figure 2.3 Kestrels accumulate prey in nest boxes or other caches. In so doing, they can optimise their hunting effort in relation to factors such as hunting success or weather conditions. This strategy enables them to avoid prolonged periods without food. Photograph by David Costantini.

Epstein 1978, 1981; Ehleringer & Rundel, 1988). Differences in isotope ratios between animals may be due to food webs differing in complexity, food webs based on different primary producers and different foraging habitats. Nitrogen isotopes indicate the trophic level of an individual (i.e. the position it occupies in a food chain). On average, there is a 3.2% enrichment of ^{15}N at each trophic level. Variation in carbon isotopes indicates different foraging habitats because plants differ in carbon isotope ratios (e.g. C3 vs. C4 plants). By looking at correlations between values of stable isotopes of both predator and prey, it is also possible to estimate the prey composition and to identify the main prey species (DeNiro & Epstein, 1981; Fry & Sherr, 1988; Tieszen & Boutton, 1988; Inger & Bearhop, 2008; Ben-David & Flaherty, 2012). Thus, analyses of stable carbon and nitrogen isotope ratios in kestrels and their potential prey might help to overcome some of the above-mentioned methodological limitations of pellet or fresh remains analyses in the estimation of diet composition. An interesting approach that has been tested on lesser kestrels analysed other isotopes (δ^{34}S, δ^2H, δ^{18}O), in addition to δ^{13}C and δ^{15}N, in feathers and claws of the same individual to assess differences among colonies living in different habitats (Morganti et al., 2016).

Figure 2.4 Western green lizards are an important prey species for kestrels in central Italy (Costantini et al., 2005, 2007c). Photograph by Gianluca Damiani. A black and white version of this figure will appear in some formats. For the colour version, refer to the plate section.

Comparisons of stable isotopes among populations also need to give careful consideration to the geographic location because local isotopic composition varies at a regional scale.

2.4   Sources of Variation in Diet Composition

2.4.1   Geography

In northern and central Europe, kestrels feed almost exclusively on voles, particularly of the genus *Microtus* (e.g. Cavé, 1967; Davies, 1975; Village, 1982a; Korpimäki, 1985a, 1985b, 1986; Masman et al., 1986; Kochanek, 1990; Śliwa & Rejt, 2006; Riegert et al., 2007b; Żmihorski & Rejt, 2007; Zombor & Tóth, 2015; Smiddy, 2017). For example, in Scotland, Village (1982a) found that about 80% of pellets contained remains of short-tailed voles (*M. agrestis*), which were 79% of small mammals caught with snap-traps in the same areas where pellets were collected. Small mammals also represent the main prey for kestrels in other regions of the world. In the Zuojia natural reserve, which is located in north-east China, rodents were the main prey items during the breeding season, comprising 93.9% of total prey items and 97.0% of total prey biomass (Geng et al., 2009).

The diet composition of kestrels living at lower latitudes, such as southern European countries, the Canary Islands, north Africa or Iran, is richer in lizards, birds and insects (e.g. Gil-Delgado et al., 1995; Fattorini et al., 1999; Baziz et al., 2001; Costantini et al., 2005, 2007c; Khaleghizadeh & Javidkar, 2006; Padilla et al., 2007; Souttou et al., 2007; Fargallo et al., 2009; Kaf et al., 2015; Navarro-López & Fargallo, 2015; Anushiravania & Roshan, 2017; Carrillo et al., 2017). In some populations, snakes are also relevant, suggesting possible local specialisations. For example, in an urban area of Algeria, Kaf et al. (2015) found that 12% of prey were represented by snakes, including the horseshoe whip snake (*Hemorrhois hippocrepis*), the viperine water snake (*Natrix maura*) and the Montpellier snake (*Malpolon monspessulanus*) (Menouar Saheb, personal communication). Small snakes can also be caught opportunistically when they are available. For example, 13 newly hatched grass snakes (*N. natrix*) were found in a single visit to a nest box located near the river Tiber, where insects and lizards were usually the main prey (personal observation). The diet composition of other kestrel species living in tropical regions is also rich in insects (primarily grasshoppers) and lizards (Madagascar kestrel: Rand, 1936; Rene de Roland et al., 2005b; Mauritius kestrel: Jones, 1984; rock kestrel: Barnard, 1987; Van Zyl, 1994; Madagascar banded kestrel: Rene de Roland et al., 2005a; Seychelles kestrel: Watson, 1992).

2.4.2   Habitat

In addition to the large-scale geographical and latitudinal variations, the relative occurrence of prey in diet also varies at a smaller scale, depending on the habitat structure of the areas where kestrels hunt (Figure 2.5). For example, habitats that are structurally more complex with vegetation and tall grass may provide cover and hinder the movement of prey. Open habitats with short grass and perching opportunities can be more convenient to spot and catch prey with a reduced effort. Consequently, kestrels appear to select the foraging sites depending on the availability of prey (e.g. Barnard, 1987) or easiness of hunting (e.g. Garratt et al., 2011).

Village (1982a) found that voles were less common, while birds, beetles and earthworms were more common in pellets collected from sheepwalk than from young plantations. Village (1990) also found that there was greater diversity of prey in farmland than in grassland areas, and the kind of land management had an impact on the diet composition. Compared to arable farmlands, the diet of kestrels in mixed farmlands was richer in earthworms and poorer in woodmice and grasshoppers. In Finland, Korpimäki (1985a) found that shrews and birds were more frequently preyed in small fields (< 10 km^2), whereas voles were taken by kestrels more frequently in medium-sized (10–50 km^2) and large fields (> 50 km^2). In central

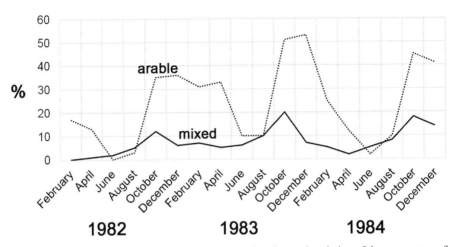

Figure 2.5 This graphic illustrates the habitat, seasonal and annual variation of the percentage of pellets containing mice regurgitated from common kestrels across English farmlands. Data collected from table 4 in Village (1990).

Italy, Piattella et al. (1999) and Salvati et al. (1999) found that birds were the most common prey in urban and suburban areas of Rome, while rodents and reptiles were more common in rural areas. Costantini et al. (2005) found that the Italian sparrow (*Passer italiae*) and Savi's pine vole (*Microtus savii*) were the main prey in cereal and cultivated fields, while the western green lizard was the main prey in open fields rich in bushes and woody patches. Casagrande et al. (2008) found that kestrels preyed on a larger number of vertebrates on grasslands than on cultivated fields. In suburban and urban environments in Algeria, birds were generally the dominant prey group (Souttou et al., 2018). Orthoptera were consumed in large numbers in farmland areas, while beetles and birds were the prey in steppe areas. Rodents represented an important component of diet in agricultural environments and also in some urban areas (Souttou et al., 2018).

2.4.3   Season

The abundance of a given prey in the environment is not constant throughout the year; rather, it can fluctuate over time, meaning that the kestrel's diet composition could also vary accordingly. This is well illustrated by lizards, which disappear from kestrels' diet during the wintertime because they go into hibernation in temperate regions. Village (1982a) reported regular seasonal fluctuations in some prey but not in others in Scotland. The frequency of voles in pellets did not show any marked changes even during the periods of increase or decrease in local abundance (Village, 1982a). Conversely to voles, birds were very common prey in the June–July period and earthworms were taken mainly in late winter or early spring and almost never in June–July (Village, 1982a). Similar seasonal variation in prey taken was also observed in other regions (e.g. England in Pettifor, 1984).

2.4.4   Year

As with seasonal fluctuations, relative abundance of prey in diet can also vary from year to year. Village (1982a) in Scotland, Korpimäki (1985b) in Finland and Casagrande et al. (2008) in northern Italy found that the abundance of voles in the diet was larger in those years when their local abundance was also larger (i.e. peak phases). However, these studies also showed that some prey were taken more than expected from their local availability as estimated by trapping (e.g. Korpimäki, 1985a, 1985b; Casagrande et al., 2008), implying a potential preference for them. It might be that the trapping methods were not adequate to provide numbers truly reflecting the local abundance of those prey. However, it might also be that other factors, such as optimisation of foraging or preferences for particular prey, contributed to generate variation in diet composition across time and environmental contexts. For example, Valkama et al. (1995) showed that kestrels may change hunting habitats in those years when a population crash of their main prey occurs.

2.4.5   Individual

The landscape diversity, in terms of habitat types that occur within the hunting area, does not necessarily reflect the diversity of prey. In the region of Campo Azálvaro in central Spain, Navarro-López and Fargallo (2015) found that the diversity of prey consumed by kestrels was lower in more heterogeneous landscapes, suggesting possible prey selection made by individual kestrels. Although several articles reported that there may be individual differences in food selection in birds, there has been surprisingly little experimental work that explicitly examined the causes and the consequences of such individual variation. For example, Brown (1969) and Gilraldeau and Lefebvre (1985) reported that individual pigeons (*Columba livia*) selected different seed types from a mixture of many seeds. Manganaro et al. (1990) observed that two neighbouring pairs of tawny owls (*Strix aluco*) had different diets: one pair preyed almost exclusively on birds, while the diet of the other pair was richer in small rodents. Differences in diet composition between neighbouring breeding pairs were also anecdotally observed in diurnal birds of prey, such as in the red-footed falcon (Purger, 1998) and in the brown goshawk *Accipiter fasciatus* (Aumann, 1988). In the American kestrel, Hart (1972) and Smallwood (1989) found marked individual differences in diet composition in free-living birds, while Duke et al. (1996) found differences among captive kestrels even in the parts of the prey eaten. A number of experiments carried out on captive American kestrels further supported the occurrence of individual differences in prey selection (Mueller, 1987).

As far as the common kestrel is concerned, Costantini et al. (2005) found differences in diet composition between pairs nesting at a short distance from each other (Figure 2.6), implying that they were sharing part of the hunting ground. Thus, these differences in diet were probably related to individual variation in feeding behaviour rather than to differences in prey availability across the environment.

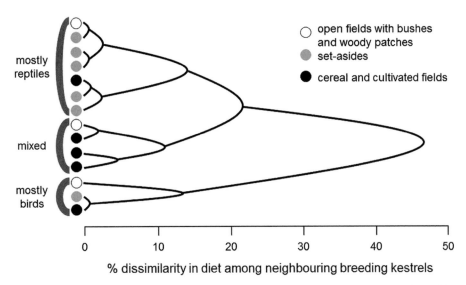

Figure 2.6 Dendrogram calculated with a cluster analysis (unweighted pair-group method with arithmetic average) based on Pianka's index (which indicates the percentage of dissimilarity among diets of different individuals). The analysis included the diet composition of neighbouring breeding kestrels, assuming that they shared the hunting grounds. The diet composition was categorised as composed mostly by reptiles, mostly by birds or mixed (absence of dominant prey). Redrawn from Costantini et al. (2005).

In the common kestrel, intraspecific competition can be excluded only within the territory, i.e. a small area (in most cases below 50–100 m radius) surrounding the nest, which is defended from intruders (e.g. Cramp & Simmons, 1980; Sommani, 1986; Village, 1990). The rest of the home range, i.e. the area in which the birds hunt actively, may overlap widely among pairs (Section 3.3; Cavé, 1967; Village, 1990). The hunting grounds shared between pairs host many potential prey species, which occur in different abundances, but also require different hunting efforts and skills in order to be caught by the kestrels. Thus, other features beyond the simple local abundance might be relevant in affecting the diet composition of kestrels. Costantini et al. (2005) also recorded regularly in specific nests some prey, such as frogs, geckos, or swifts, which were rarely found in other nests.

The available evidence suggests that individual food preferences occur in many bird species that are considered as food generalists, including the common kestrel. These individual differences in feeding behaviour might be interpreted as a component of individual behavioural profiles or personalities, which are defined as consistent and non-random differences in how individuals behave across time and contexts (Carere & Maestripieri, 2013). Thus, individuals may show less feeding plasticity than expected, contrasting the classic view of a purely generalist strategy that would enable kestrels to maximise exploitation of food resources available in the environment. In the case study reported by Costantini et al. (2005), the kestrels, whose diet was monitored for several years, always succeeded in producing viable offspring, thereby suggesting that their peculiar individual feeding habits were not maladaptive. However, the extent to which individual feeding habits carry costs for reproductive fitness might depend on the local environmental conditions. For example, when there is food shortage (e.g. during the wintertime), it might not pay off to be specialised feeders if the preferred prey occur in limited supply.

Another question, then, is how such individual differences in feeding behaviour may evolve and be maintained within a population. Temporal variation in the prey available across the breeding season could have favoured the development and main-tenance of individual feeding habits in kestrels (Costantini et al., 2005). Genetic analyses carried out on the same population studied by Costantini et al. (2005) suggested the occurrence of assortative mating-by-time and the possibility of adaptive divergence between early and late breeders (Casagrande et al., 2006a). Thus, birds breeding earlier in the season might have specialised on prey species different from those of late breeders. Although this temporal mechanism might contribute to shape individual variation in feeding habits, other mechanisms are certainly involved because individual differences in feeding habits also occurred in pairs breeding at the same time of the breeding season (Costantini et al., 2005). For example, the occurrence of different individual feeding preferences might reduce the intraspecific competition for food, which would be particularly advantageous in times of food shortage. Frequencies of both specialised and generalist individuals within the popula-tion would eventually depend on this balance between costs and benefits.

The development of a *search image* for a certain prey is common in predators. *Food imprinting* could represent a way of transmission from parents to offspring of a familial preference for a specific prey through the development of a specific search image (Tinbergen, 1960; Mueller, 1971). In so doing, food imprinting could favour the maintenance of a 'cultural' propensity to develop individual feeding habits in the population (Stokes, 1971; Allen & Clark, 2005).

For a long time, among-individual differences in food preferences have been neglected as biologically relevant because they were considered extremes of popula-tion means with a merely descriptive value. Costantini et al. (2005) suggested another way to look at these individual differences by considering feeding behaviour as part of a behavioural profile, which could be linked to other traits of avian personalities (Carere & Maestripieri, 2013). These unanswered questions open novel avenues for research into the feeding ecology of kestrels and other bird species.

2.4.6   Sex and Age

Female kestrels are around 20% larger than male kestrels. These differences in body size might imply that females and males also differ in the propensity or capability of catching prey of different size. Village (1990) found that males took more insects than females in both mixed and arable farmlands. Sexual differences in diet might also be explained to some degree by separation of hunting habitats between sexes. Meyer and Balgooyen (1987) found that male American kestrels caught more insects than did females. These sexual differences in diet were due to a different availability of prey in the respective hunting areas of males and females, with larger prey being present in the females' hunting habitats (Meyer & Balgooyen, 1987). Although competitive exclusion is important in the American kestrel (Ardia & Bildstein, 1997), sexual differences in diet might imply that males and females use the size of prey as a further clue to select the habitat where to hunt. Differences in hunting behaviour between males and females are also relevant in other kestrel species (Section 2.7.1). Differences in diet were also observed between juvenile and adult kestrels (Village, 1990). Generally, young kestrels feed more on insects than do adults, probably because of their limited hunting experience.

2.5 Food Competition

Intra- and interspecific competition for food is an important evolutionary force that shapes the niches of coexisting species that feed on the same prey to a larger extent. In many areas of central and northern Europe, common kestrels share their breeding environments with long-eared owls (*Asio otus*). Kestrels and owls have similar feeding ecologies, which makes the comparison of their diets an excellent system to understand the role of competition in shaping the diet of kestrels. A study carried out in Finland found that the similarity in diet composition between kestrels and long-eared owls was lower in years with low abundance of voles, and neighbouring breeding pairs of both species showed less overlap in diet composition than non-neighbouring breeding pairs as expected by the competition theory (Korpimäki, 1987). Importantly, shared and unshared breeding habitats were similar in composition and local availability of prey (Korpimäki, 1987), suggesting that interspecific competition might have been a significant driver of diet composition.

2.6 Kleptoparasitism

Kleptoparasitism is a foraging tactic in which one animal steals prey or other food from another that has caught or collected it. Optimal foraging theory predicts that an individual should be kleptoparasitic when the net energy derived from this strategy exceeds the net energy gained from searching for food itself (Brockmann & Barnard, 1979). Kleptoparasitism might also be an energy-saving strategy when conditions for hunting are not optimal. Fritz (1998) actually found that kleptoparasitism was more frequent when wind speed was too weak to sustain flight-hunting. However, many other aspects independent from energy intake, such as muscle inflammation or molecular oxidative damage due to intense foraging activity (Chapter 7), might also be important. Other hypotheses considered kleptoparasitism as either a strategy of dominance or as a way of signalling individual quality to others in order to displace potential competitors from better-quality feeding sites (e.g. Bautista et al., 1995, 1998).

The common kestrel can be either victim or executioner of kleptoparasitism (Brockmann & Barnard, 1979). Several species of birds of prey, such as the Eurasian sparrowhawk (*Accipiter nisus*), the short-eared owl (*Asio flammeus*), or the peregrine falcon, have been observed to steal prey from kestrels (Brockmann & Barnard, 1979). Large passerines may also be important kleptoparasites of kestrels. In a three-year field study carried out in Poland, Kitowski (2005) observed that kestrels were subject to kleptoparasitism attempts by numerous bird species, including the hooded crow (*Corvus corone*), the Eurasian magpie, the western jackdaw (*Coloeus monedula*), the rook (*Corvus frugilegus*) and the raven (*Corvus corax*). Kleptoparasitism may have important consequences because robbery of prey can account for a non-negligible loss in biomass of prey captured by kestrels. Kitowski (2005) found that the overall number of prey lost because of kleptoparasitism actually constituted 19.3% of all prey caught in the foraging area and 21.3% of the total biomass of prey captured. One important consequence of such loss of food was that some kestrels abandoned the area. In contrast, kestrels that did not leave the area restarted hunting within a time frame of around 3–25 minutes after being kleptoparasitised, with females waiting longer than males before hunting again. These observations would suggest that kleptoparasitised individuals might face higher metabolic demands to sustain foraging activity.

Kestrels may also kleptoparasitise other birds (Brockmann & Barnard, 1979) and, apparently, even mammals (Elliott, 1971). For example, kestrels have been observed rolling over in flight and snatching a small mammal from a flying barn owl (*Tyto alba*) (Everett, 1968) or from short-eared owls (Clegg & Henderson, 1971). Fritz (1998) recorded 25 attempts made by kestrels to steal prey from short-eared owls in 16 afternoons of observations.

Overall, these studies on kleptoparasitism are largely anecdotal, consisting mainly of observations and personal experiences. These results raise interesting questions in relation to the costs incurred by kleptoparasitised kestrels. Although kestrels seem to occasionally kleptoparasitise other species, the causes and consequences for doing kleptoparasitism are currently unknown.

2.7   Foraging Strategies

2.7.1   Hunting Behaviour

Kestrels travel from a home base to a distant foraging location and carry single prey to the nest or perch. Most observations of the hunting behaviour of kestrels have been focused on determining general patterns with little attention to individual variation and the development of this behaviour (e.g. learning of hunting mode during the post-fledging period). In general, three main hunting modes can be recognised (Figure 2.7): wind-hovering, soaring, and perching (Village, 1990).

Wind-hovering is the most familiar feature of the common kestrel. Birds remain stuck above a point on the ground (i.e. zero ground speed) by beating the wings continuously into the wind and, while wind-hovering, they patrol the ground looking for prey. Thus, wind-hovering includes both flapping and gliding flight during the stationary sessions of flight-hunting (Videler et al., 1983). Although wind-hovering is

Figure 2.7 This image shows a kestrel while wind-hovering (above) and another one that caught an insect while on the wing (below). Photographs by David Costantini.

typical of many kestrel species, this hunting technique is uncommon in grey kestrels, the Madagascar kestrel or the fox kestrel (Gaymer, 1967; Boyce & White, 1987; Londei, 2002).

Soaring consists of slow movements in the air with little or no wing movements and is mostly restricted to warm periods with thermal air lifts. Perching indicates sitting above the ground on a perch with a certain degree of view of the environment and the potential prey moving in it. Kestrels may also catch prey while sitting on the ground. This method of hunting is mostly used to catch invertebrates like insects or earthworms. This variation in hunting modes gives the kestrels enough flexibility to adjust their way of hunting to their current state (or health) or to the prevailing weather conditions.

In a three-year study, Rijnsdorp et al. (1981) observed that wind-hovering yielded 76% of all prey and that it was drastically reduced or even impeded by rain, fog and wind speeds below 4 m s^{-1} and above 12 m s^{-1}. They also found that 97% of prey taken by wind-hovering kestrels were small mammals and 3% were birds, while bird percentages were 34% and 100% for perching and soaring, respectively. Motivation to hunt was also important; hunting frequency increased with the time elapsed since the last prey was taken or following unsuccessful strikes (i.e. attempt to catch a prey). Further work on kestrels in the Netherlands showed that the number of prey caught varied largely among hunting techniques and that wind-hovering mainly reflected the seasonal variation in prey abundance (Masman et al., 1988b). Wind-hovering yielded, on average, 2.2 small mammals per hour in winter and 4.7 in summer, while perch-hunting yielded 0.3 and 0.1 small mammals in winter and summer, respectively. Thus, overall, perch-hunting in summer yielded about 6% of prey, while in winter it yielded about 24%. It has been proposed that minimisation of energy expenditure might be one reason for the winter hunting behaviour, while in summer minimisation of foraging time and time of nutrient assimilation were also probably important (Masman et al., 1988b). However, minimisation of other costs associated with intense flight activity (e.g. muscle inflammation, molecular oxidative damage) might also be important in determining the selective pressures on the hunting strategies.

Several authors also observed that the frequency of wind-hovering increases from winter to summer and then decreases after the young become independent (e.g. Shrubb, 1982; Pettifor, 1983; Masman et al., 1986; Village, 1990). During the nestling-rearing period, adults need to catch more food, which makes wind-hovering preferable because it enables kestrels to catch food quickly enough to meet the food demands of the family.

Habitat composition is one important environmental factor that affects the hunting behaviour of kestrels. For example, a shortage of suitable perches would inevitably reduce the perch-hunting behaviour. To test this prediction experimentally, Lihu et al. (2007) increased the potential perch sites in one area and kept another area as control. They found that in the test area with more perch sites, kestrels hunted 77% of the total hunting with the technique of perch-hunting, while kestrels hunted only with the technique of wind-hovering in the control area.

Mobbing of a predator by potential prey species is also important in determining its hunting behaviour. Pettifor (1990) found that wind-hovering kestrels were mobbed more than expected compared with those that were perch-hunting. Importantly, both wind-hovering and perch-hunting kestrels flew significantly further between their foraging positions when they were mobbed than when they were not mobbed. Moreover, on average, mobbing resulted in wind-hovering kestrels flying 6.8 times and perch-hunting kestrels flying 2.7 times the mean distances moved by non-mobbed birds (Pettifor, 1990). Thus, mobbing might significantly increase the demands of wind-hovering.

Males and females may also differ in several aspects of their hunting behaviour. Such differences have been recorded in several kestrel species, such as the Australian kestrel (Genelly, 1978; Paull, 1991), the common kestrel (Shrubb, 1982; Masman et al., 1986), the greater kestrel (Hustler, 1983; Kemp, 1995), the American kestrel (Meyer & Balgooyen, 1987; Toland, 1987) and the Mauritius kestrel (Temple, 1987). For example, Kemp (1995) observed that male greater kestrels made more strikes than females into tall grasslands (30 vs 14%), more from perches (77 vs 49%), and more (23 vs 6%) but with lower success (31 vs 66%) at vertebrates due to more misses at birds. Females made more strikes into short grasslands, more at invertebrates, and with more success from hovers (96 vs 66%). Kemp (1995) also recorded sexual differences in perch heights and in hunting behaviour at different times of the day. However, the reasons for these differences between male and female kestrels currently remain untested experimentally.

2.7.2 Caching Behaviour

Once a kestrel has caught a prey, it may bring it to the nest or eat it partially or entirely. Kestrels can also cache prey for later retrieval (e.g. Leaver, 1951; Clegg, 1971; Parker, 1977; Rijnsdorp et al., 1981; personal observations), indicating that they make a choice of when to eat. Caching is also common in other birds of prey, including other kestrel species (American kestrel: Collopy, 1977; Rudolf, 1982).

Prey are usually cached in grass clumps, shrubs, or fence posts. However, incubating females may also cache prey near the nest. For example, some females breeding inside nest boxes attached to pylons of utility lines were observed to cache prey on top of the nest box or on the pylon itself (personal observations). It is thought that the caching behaviour is relevant for optimisation of hunting behaviour and food intake, thus it should be considered when analysing the temporal adjustment of foraging behaviour. For example, caching could provide a fail-safe to avoid the situation where failure in hunting or bad weather conditions leave kestrels without food for some time.

Rijnsdorp et al. (1981) suggested that the caching behaviour enables kestrels to maintain high rates of capture, uninterrupted by inactive behaviour following excessive feeding, at times when prey are abundant, and in order to have a late afternoon meal before a cold night. The hypothesis that caching enables kestrels to resume hunting without delay does not seem to be well supported because kestrels may eat most prey within few minutes (Masman et al., 1986; Village, 1990). Caching may also enable incubating females to dampen a reduction in prey delivery by their mates (Wiebe et al., 2000). American kestrels deprived of food for long periods cache more prey when they are later given surplus food (Mueller, 1974). Thus, females may use caching strategically as an anticipatory response to forthcoming periods of food shortage.

Rijnsdorp et al. (1981) observed that the caching behaviour is prevalent during winter and is done at any time of the day, but rarely in the last two hours before dusk. Indeed, most cached prey were seen to be retrieved just before dusk, as was also observed in American kestrels (Collopy, 1977). Rijnsdorp et al. (1981) further observed that voles cached and retrieved at the end of the day were often larger than voles eaten immediately after capture, as previously described in captive American kestrels (Mueller, 1973), or delivered to the female (Masman et al., 1986). These observations also raise the intriguing question of which cognitive abilities (e.g. episodic-like memory) enable kestrels to make decisions about caching and recovering their prey.

2.7.3   Association with Other Species

Associations with other species can provide kestrels with further foraging opportunities, but this topic has been poorly explored so far. For example, rock kestrels were found to follow groups of chacma baboons (*Papio ursinus*) in Namibia because, in so doing, they could easily catch insects that flew away due to their disturbance by baboons (Figure 2.8; King & Cowlishaw, 2009). Anecdotal observations of the foraging behaviour of Madagascar banded kestrels indicated that they might associate with sickle-billed vangas (*Falculea palliata*) while looking for food because, as with rock kestrels, they were observed to easily catch insects that flew away due to their disturbance by the vangas (Tingay & Gilbert, 2000).

Figure 2.8 Schematic representation of foraging opportunities derived by kestrels from their association with baboon groups. Reprinted from King and Cowlishaw (2009) with permission from John Wiley and Sons and from Springer Nature.

2.7.4   Novel Methods to Characterise Remotely the Foraging Behaviour

The quantification of activity patterns of kestrels has not only been time-consuming but also sometimes logistically challenging to carry out. This means that many studies are limited to record the activity of a moderate number of individuals for short durations. In recent years, the advent of miniaturised tracking technologies has made the recording of activity patterns easier to carry out even on a large number of individuals for long periods. Accelerometers and GPS data-loggers are playing a central role in advancing movement ecology. Accelerometers are small devices that can be attached to an animal's body and take recordings of body acceleration remotely (proxy of body motion; e.g. Nathan et al., 2012; Hammond et al., 2016). Accelerometer data can be used to derive daily activity budgets in order to determine when an animal moves (Yoda et al., 1999; Grünewälder et al., 2012; Costantini et al., 2017b). If accelerometer and viewer-observed behavioural data are collected simultaneously, it is possible to determine which body acceleration values are generated by different behaviours (e.g. gliding, flapping, walking). In so doing, accelerometer data can also be used to characterise which behaviours an individual was doing and for how long and at which time of the day these behaviours were being performed (Sakamoto et al., 2009; Nathan et al., 2012; Graf et al., 2015), and to estimate the energy expenditure associated with different behaviours (Wilson et al., 2006; Qasem et al., 2012; Jeanniard-du-Dot et al., 2017). A recent study on lesser kestrels showed that accelerometers can be successfully deployed on these medium-size raptors and that the data collected can be used to discriminate among different behaviours with a low classification error (Hernández-Pliego et al., 2017). Another interesting approach has been proposed by Rodríguez et al. (2012). Immediately after recovering the data-loggers from lesser kestrels that returned from a foraging trip, the authors loaded the spatiotemporal information gathered from the GPS tags into an unnamed aerial system (UAS, i.e. a drone), which repeated the same flight path flown by the birds. The UAS had an onboard camera that enabled documentation in quasi-real time of the habitat visited by the birds.

2.8   Visual Cues for Prey Selection

Kestrels may rely on several cues, such as body size, body colouration or activity level, to target their prey (Bryan, 1984; Sarno & Gubanich, 1995; Costantini et al., 2007c). Viitala et al. (1995) proposed that common kestrels are attracted to vole urine and faeces marks because they both reflect ultraviolet light, a range of the light spectrum that birds can see. Kestrels appear to be able to differentiate between vole species and their reproductive statuses on the basis of scent marks, which might have important consequences for prey choice and selection of both hunting and breeding areas (Koivula et al., 1999). However, later work on the ocular media transmittance suggested that vole urine does not seem to provide a reliable visual cue under natural circumstances where signals might be (i) confounded by urine from other mammals,

(ii) destroyed by rain and wind and (iii) seen from a distance where the low spatial resolution of chromatic vision has to be considered (Lind et al., 2013). Thus, the role of UV vision in prey selection remains unclear.

Attributes of prey are certainly important; however, it is unclear which ones drive selective predation or if their interaction amplifies the signal that kestrels may perceive. Costantini et al. (2007c) found that male green lizards are significantly more preyed on than female green lizards in central Italy. In contrast, a sex-biased predation was not detected in two smaller lizard species (*Podarcis muralis* and *P. sicula*) in the same study areas (Costantini & Dell'Omo, 2010). As male green lizards are slightly larger than females, the body size might be less relevant than other factors (e.g. higher activity of males in open areas, nuptial colouration) in determining their higher risk of being predated. It has been suggested that the higher activity of males might be the main reason for the higher predation of male voles in Finland (Korpimäki, 1985b).

2.9   Kestrels as Secondary Seed Dispersers

The role of birds of prey as seed dispersers has been poorly investigated so far. Raptors that feed on frugivorous prey might act as relevant secondary seed dispersers. In the Canary Islands, common kestrels feed largely on frugivorous lizards of the genus *Gallotia*. Kestrels dismember the lizards and discard the digestive tract, which contains the seeds of several plant species (Padilla & Nogales, 2009). In so doing, most seeds only undergo a partial digestion process within the lizard's digestive tract. For this reason, seeds retained a similar germination capacity to that of seeds collected from plants and even higher than that of seeds that underwent the whole digestion process in lizards (Padilla & Nogales, 2009). Further work showed that kestrels from Canary Islands might disperse seeds of many plant species at distances up to about 1 km from the site where the lizard was caught (Padilla et al., 2012). These results suggested that kestrels might have an important role as secondary seed dispersers for some plant species in insular ecosystems.

2.10   Conclusions

In conclusion, the common kestrel can be considered as a generalist species. However, large within-species variation in diet composition occurs. Individual differences in feeding behaviour are, for example, an important but little explored source of this within-species variation in feeding ecology. The biological meaning of this individual variation remains an open question.

This chapter showed that most of what we know about the feeding ecology of kestrels comes from research carried out in central and northern Europe. The findings of these studies might not apply to kestrels living at lower latitudes because southern kestrels have a broader trophic niche, feed more on reptiles and insects and have more sedentary habits. Moreover, cycles of prey populations vary across regions in terms of amplitude, dynamics and predictability. We need to know more about the feeding ecology of kestrels (including its numerous subspecies) in other geographic regions in order to draw general patterns. In so doing, it will be important to use a more experimental approach because much of what we know is based on correlational studies.

Another research limitation has been the difficulty in obtaining unbiased estimates of prey composition. Analysis of stable isotope ratios of carbon and nitrogen would certainly enable us to overcome many of those limitations. Finally, we need research on the interactions among endogenous circadian control of individual daily routines, individual experience/feeding habits and fluctuations of prey in determining foraging behaviour. Research in this direction would foster our understanding of how individual kestrels optimise foraging in relation to other important daily activities across the year.

3   Habitat Use

3.1   Chapter Summary

The size of home ranges of common kestrels can vary dramatically among individuals. Within the home range, each individual kestrel defends a small area around the nest from intruders, which is referred to as territory. Home ranges are dynamic because their size varies across the year. Also, kestrels do not use homogeneously all habitat types within their home range, but show preferences for certain habitats. The first GPS-tracking study reported in this chapter supports early findings, but opens new avenues to improve data collection on habitat use and home-range size estimate. Finally, this chapter shows that the urban environment might not be a particularly suitable home range for kestrels because the available evidence suggests that urban kestrels have a poorer reproductive performance than rural kestrels.

3.2   Introduction

Understanding the habitat use of wildlife and its consequences on population viability are research priorities in ecology and conservation science. Habitats provide a number of resources and conditions that influence the survival and reproduction of animals. In turn, animals may optimise habitat use by selecting those habitat types that have richer food availability, better microclimatic conditions or lower competition. However, environmental clues used by animals to select given habitats might sometimes be misleading, leading them to fall into an *ecological trap* (settlement in poor-quality habitats even though alternative high-quality habitats are available).

The habitat use of common kestrels was studied in a few areas across their entire geographic range. These studies relied on methods that suffer from several drawbacks, so our understanding of the causes and consequences of individual variation in habitat use is still limited. The advent of new tracking technologies (e.g. GPS data-loggers) has revolutionised the research on animal movements (Chapter 9) and habitat use. Their application in ecological research on kestrels will certainly open new avenues to better understand how kestrels use their habitats and why they do so (Section 3.4). This has profound conservation implications because detailed data on habitat use are fundamental to guide stakeholders and policy makers to implement evidence-based landscape planning to limit anthropogenic impacts on kestrels.

3.3   Home Range

Animals live in a given area, where they perform their day-to-day activities, such as breeding or foraging. Such an area is called the home range. Some species may exclude conspecific individuals from using part of it; this defended area is called the territory. Quantification of home range and territory sizes were classically performed by marking kestrels (e.g. wing tags, radio transmitters) so that they could be individually recognised in the field. This approach requires intensive field observations of animals and can, therefore, be subject to bias because it is not possible to observe the animals continuously. Moreover, conversely to GPS data-loggers, the error in the identification of locations is quite variable across estimates, which makes comparison of studies complicated (see appendix 2 in Village, 1990). The deployment of GPS data-loggers on kestrels would overcome many limitations of other tracking methods, enabling accurate measurement of the size of the home range. However, so far, the only application of GPS data-loggers on common kestrels is that presented in Section 3.4.

The size of the home range is generally influenced by several factors, such as prey availability and competition, which are also connected to each other to some degree. By following individually marked birds, Village (1990) observed that kestrels rarely use all parts of their home range with equal intensity, suggesting that it might not be straightforward to determine whether an individual enlarged its home range when it moved into a new area. Village (1982b) also observed that kestrels rapidly expanded their home ranges as soon as neighbours left the area for dispersing or migrating. Shrinking of home ranges occurred again in the spring, when the arrival of migrants increased local density and, consequently, territorial displays and fights. These results pointed to a significant role of competition in determining the size of the home range.

The size of home ranges calculated for kestrels inhabiting the Scottish grasslands and the English farmlands using the minimum convex polygon method varied from a mean value of 0.74–4.03 km^2 in autumn (October and November), 1.37–4.87 km^2 in winter (December–March), and 3.11–12.86 km^2 in summer (April–July) (Figure 3.1). The overlap among home ranges varied from 5% to 20% in autumn, 6% to 30% in winter, and 4% to 49% in summer (Figure 3.1). The percentage of overlap increased with distance from the nest; overlap was also greater in years with larger abundance of prey, when the local density of kestrels was also higher, determining clumping of nests (Village, 1983). Most kestrel activity was recorded within 1 km of the nest, with a few excursions up to 4 km during the late nestling stage (Village, 1990). The greater overlap in summer than in winter might be due to a higher local density owing to the arrival of migrants. However, kestrels can also change their hunting behaviour across the year, doing more perch-hunting in winter and flight-hunting in summer. Thus, kestrels might have been less aggressive against neighbours during summer because they needed to overlap their home ranges due to their increased flight activity. Also, the minimum convex polygon is a method that can overestimate the home range. It is the area obtained when all outer recorded locations are joined to form a convex polygon, so that it is unknown if kestrels used the whole area within the polygon.

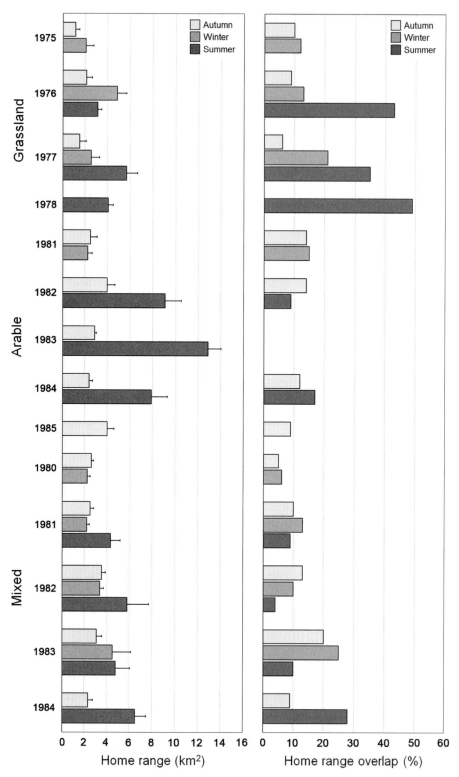

Figure 3.1 Size of home ranges of common kestrels in Scotland (grassland) and England (arable and mixed lands). For each range was calculated the minimum polygon area (MPA) by connecting the outer locations to form a convex polygon. The MPA was then corrected for the sample size. Data from Village (1982b, 1990).

In northern Italy near the city of Parma, home ranges of male kestrels recorded by Casagrande et al. (2008) varied from 0.28 to 1.42 km². These values are similar to those recorded by Beichle (1980) in the city of Kiel in the north of Germany (0.9–3.1 km²) or Boileau et al. (2006) in marshlands of western France (0.5–3.8 km²), but are rather small if compared to those recorded by Village (1990), although both studies used radio transmitters to record the movements of birds and used the same methods to calculate the home-range size. Riegert et al. (2007a) studied the home ranges of 34 male kestrels in a small city (40 km²) in southern Bohemia (Czech Republic). Using the minimum convex polygon method, they determined that the home range varied from 0.8 to 25.0 km² with an average of 7.2 km² with little home-range overlap (< 10%) among males breeding at the periphery of the city. In the city centre, there was greater home-range overlap (30–50%) and the home-range size itself was larger than that of males breeding at the periphery. Also, the home range of some city centre males overlapped much of the home range of periphery males.

Another limitation of the above-reviewed studies is that it is hard to determine precisely whether individual kestrels had any preferences for given hunting grounds. Kestrels do not use the whole home range homogeneously, but tend to use some parts of it more than others (e.g. Village, 1990). In northern Italy, Casagrande et al. (2008) observed that the grasslands were the most-used habitats during both flight- and perch-hunting, while controlling for habitat available within each home range. They also recorded some differences in habitat use between the two hunting activities. Riverbanks were mainly used during the flight-hunting, while the set-asides were the least-used habitats among those available. Conversely, set-asides were the most used habitat during the perch-hunting. The utilisation of habitats was significantly explained by the local availability of voles.

3.4   GPS-tracking of Common Kestrels

The use of GPS data-loggers enables the limitations of the above-reviewed articles to be overcome. They enable to record almost continuously the movements and the locations of individual kestrels without the need of carrying out visual observations. The plot of data on a land-cover-/use map enables quantification with precision of the time spent by the birds in each habitat type. With this information, it is possible to determine whether birds have any habitat preferences by a statistical comparison of fixes in each habitat in relation to the available habitat (e.g. compositional analysis). It is also possible to determine whether the birds were flying or perching using the speed of the animal recorded by the GPS data-logger. Moreover, the measurement error is lower and better standardised than that of visual observations or radio transmitters. Despite all these advantages, so far there has not been any application of GPS tracking to common kestrels.

In 2019, we deployed for the first time GPS data-loggers (Technosmart, Italy) on two female kestrels that were breeding close to each other (about 300 m apart) and on both members of a pair. The two females were breeding in the countryside of Rome, while the

Figure 3.2 Deployment of GPS data-loggers on a male (top) and a female (bottom) common kestrel. Loggers were deployed by backpack (male) and leg-loop (female) harnesses and were equipped with a small solar-panel. Photographs by Gianluca Damiani (male kestrel) and David Costantini (female kestrel).

pair was breeding in a suburban area of Rome. Loggers were deployed by backpack (male) and leg-loop (female) harnesses as shown in Figure 3.2. Each logger was equipped with a small solar-panel to use sunlight to recharge the battery. Data were downloaded remotely through radio communication. The total mass of the equipment did not exceed 3% of the kestrels' body masses, thus it was within the generally recommended limits for flying animals. We configured the GPS devices to the sampling frequency of 1 fix/30 min. The loggers provided spatial location, altitude, and instantaneous speed, which enabled us to determine whether the kestrel was flying or perching.

The size of the home ranges was calculated by the kernel density estimation (95%). The average home range of females during the breeding season was

Table 3.1 Average home-range size quantified with the kernel density estimation (95%) using data collected from GPS tracking of common kestrels in central Italy. n. fix = number of times that the GPS recorded the position of the bird.

Sex	Area (ha, km^2) (n. fix)
Female	54.9, 0.549 (402)
Female	36.6, 0.366 (451)
Female	77.9, 0.779 (528)
Male	143.0, 1.43 (1,283)

Table 3.2 Home-range size before and after the middle stage of the nestling-rearing period quantified with the kernel density estimation (95%) using data collected from GPS tracking of common kestrels in central Italy. n. fix = number of times that the GPS recorded the position of the bird.

Sex	Area before middle stage of the nestling rearing-period (ha, km^2) (n. fix)	Area after middle stage of the nestling-rearing period (ha, km^2) (n. fix)
Female	4.6, 0.046 (113)	78.6, 0.786 (289)
Female	2.1, 0.021 (201)	70.4, 0.704 (250)
Female	33.4, 0.334 (268)	104.8, 1.048 (260)
Male	124.0, 1.24 (859)	146.7, 1.467 (424)

0.56 km^2, while it was 0.78 km^2 for the male (Table 3.1). Females tended to forage close to the nest and never flew beyond a distance of 1.5 km from the nest. The male made a few long flights, up to about 10 km from the nest. The home ranges of females were smaller before than after the middle stage of the nestling-rearing period (Table 3.2). The overlap between the females' home ranges was over all 2.4 ha, which corresponded to 4.4% and 6.6% of the home range of each female, respectively (Figure 3.3). We did not detect any overlap between the home ranges of the two females before the middle stage of the nestling-rearing period. We recorded an overlap of 12.2 ha between the home ranges of the two females nesting close to each other after the middle stage of the nestling-rearing period; this overlap between home ranges corresponded to 15.5% of one female's home range and 17.3% of the other female's home range, respectively.

As far as the pair is concerned, the male had a home range much larger than that of the female before the middle stage of the nestling-rearing period; the home range of the female became larger with the season (Table 3.2). The overlap between mates' home ranges was overall of 51.8 ha, which corresponded to 66.5% and 36.2% of the female's and of the male's home range, respectively (Figure 3.4). The overlap between home ranges of mates was 22.4 ha before the middle stage of the nestling-rearing period and corresponded to 67.1% of the

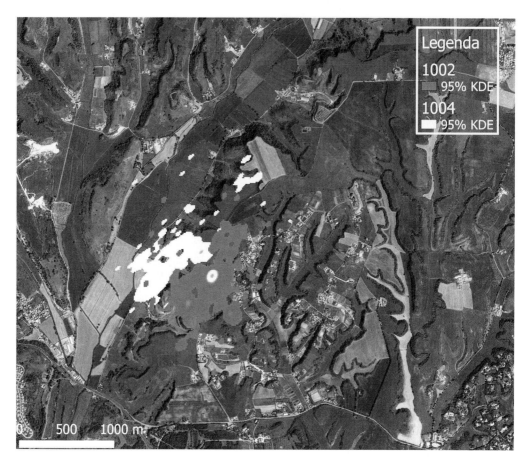

Figure 3.3 Home ranges of two female kestrels breeding close to each other (around 300 m) estimated by the fixed kernel method. A black and white version of this figure will appear in some formats. For the colour version, refer to the plate section.

female's home range and 18.0% of the male's home range, respectively. The overlap between home ranges became 66.2 ha after the middle stage of the nestling-rearing period and corresponded to 63.2% of the female's home range and 45.1% of the male's home range, respectively.

We then analysed the fixes with a recorded speed below 3 km h^{-1}, which indicate roosting. We identified a total of 47 roosts: 53% were trees, 38% were pylons of electricity lines and 9% were buildings (these data refer to the pair, whose nest box was located on a building). The average distance of the preferred perch from the nest was about 360 m. For the two female kestrels breeding in the mixed environment of cultivated and Mediterranean scrub, perches were usually located along rows of trees or pylons at the edge of country roads. As far as the couple breeding in the urban environment is concerned, apart from trees close to the nest, the male also used to perch on a silo, located about 600 m from the nest, during the night.

Figure 3.4 Home ranges of a kestrel pair estimated by the fixed kernel method. A black and white version of this figure will appear in some formats. For the colour version, refer to the plate section.

3.5   The Kestrel in the Urban Environment

Increasing urbanisation of rural landscapes has created new challenges for wildlife. The urban environment can induce a number of changes in animal behaviour, physiology and life history. The kestrel is a common inhabitant of cities; it was already present in some European cities (e.g. London, Paris) in the second half of the nineteenth century (e.g. Montier, 1977; Malher & Lesaffre, 2007; Fraissinet, 2008) and its density in urban areas has strongly increased in the last few decades. However, it is still unclear whether the urban environment represents a suitable habitat or an ecological trap for the species.

A number of studies carried out in the Czech Republic, Poland and central Italy found that the urban kestrel populations had higher reproductive rates than the rural ones (Pikula et al., 1984; Plesník, 1990; Rejt, 2001; Salvati, 2002). However, the comparison between urban and rural kestrels in those studies did not consider the potential confounding effect that the nest type may have on the estimate of breeding success, as the majority of rural nests were open-type natural nests and the urban pairs

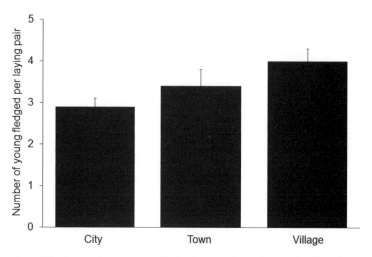

Figure 3.5 The number of young fledged per laying pair was significantly greater in kestrels breeding in buildings in rural villages than that of kestrels breeding in cities. The study was carried out in Israel. Cities were highly developed and densely populated (> 100,000 inhabitants); towns had smaller human populations (11,000–25,000 inhabitants) and smaller buildings that were more widely spaced than in cities but denser than in villages; villages had very small populations (< 700 inhabitants), few buildings, many trees, bushes and fields and agricultural areas. Data from Charter et al. (2007a).

bred in buildings (i.e. more protected cavity-like nests). Reproductive performance was similar among kestrels breeding in nest boxes in the city centre, in a mixed zone, or in the outskirts of Berlin (Kübler et al., 2005). The number of young fledged per breeding pair and the percentage of pairs that successfully fledged at least one young were lower in cities (2.9, 78.2%) than in villages (4.0, 92.3%) in Israel (Charter et al., 2007a; Figure 3.5). The majority of urban kestrels in Israel were reported breeding in flowerpots on windowsills, whereas in the three European studies the birds bred mainly in cavity nests, such as in vent holes, cracks in walls and attics and only occasionally on windowsills. Although cavities in historic buildings might offer better protection for the brood, this does not necessarily translate in higher reproductive success if the urban environment is of low quality. In Vienna, the use of urban habitats was associated with higher nest failure, partly associated with predation and nest desertion, and with significantly lower hatching rates and smaller fledged broods, possibly due to lower food quality (Sumasgutner et al., 2014c). Given that the first returning kestrels from wintering grounds selected breeding sites in the city, it might be concluded that kestrels relied on environmental clues that did not reflect the real quality of the breeding environment.

The selection of urban habitats or competitive exclusion of given genotypes might also lead to some degree of genetic isolation and differentiation as a result. Rutkowski et al. (2006) found moderate and significant differentiation among kestrels inhabiting rural, suburban and urban areas of Warsaw. This might indicate some degree of reproductive isolation; however, founder events (e.g. given genotypes colonised the city or were favourably selected in the urban environment) and genetic drift could be alternative explanations for this genetic differentiation between rural and urban kestrels in Warsaw. In southern Bohemia, however, rural and urban kestrel populations showed a similar degree of genetic polymorphism (Riegert et al., 2010).

A recent review and meta-analysis on the effects of urbanisation on the breeding performance of raptors concluded that the common kestrel is negatively affected by urbanisation (Kettel et al., 2018). Similarly, both lesser and American kestrels also appear to have a lower breeding performance in urban habitats (e.g. Tella et al., 1996b; Strasser & Heath, 2013). For example, Strasser and Heath (2013) found that proximity to large, busy roads and developed areas affected negatively reproduction of American kestrels by causing increased stress hormones that probably promoted nest abandonment. Nest site availability in the urban habitat for common kestrels might be reduced in the years to come by the widespread renovation of inner-city buildings, which has already caused a drastic decline of available nest sites in several European cities, such as Paris (Malher et al., 2010), Bardejov (Mikula et al., 2013) and Vienna (Sumasgutner et al., 2014a).

3.6   Conclusions

The study of habitat use and the quantification of home range were classically done by marking kestrels using wing tags or radio transmitters, so that they could be individually recognised in the field. These approaches enabled identification of food availability or competition as two potentially important drivers of habitat use and preferences of kestrels. However, these approaches suffer several drawbacks that can be overcome by relying on GPS-tracking, as illustrated in Section 3.4.

The kestrel has become a common inhabitant of the urban environment. However, this novel habitat might be suboptimal, raising the question of whether an ecological trap is ongoing. The available evidence suggests that rural habitats are generally better for kestrels than are urban habitats. As evident by the studies reviewed in this chapter, the work done so far sometimes suffered a lack of standardisation of methods by which data were collected in both rural and urban environments. Also, a direct comparison between rural and urban studies of breeding performance of common kestrels is not always feasible because many rural studies relied on the use of nest boxes, which offer better protection and space for laying more eggs than do the cavities of old city buildings. Thus, we need more comparative studies with a better standardisation of data collection protocols before we can definitely conclude whether the common kestrel is a successful urban exploiter.

4 Breeding Density and Nest Site Selection

4.1 Chapter Summary

The number of birds breeding in a given area (breeding density) is affected by several biotic and abiotic factors. The availability of suitable nesting sites plays a major role in determining the size of the local breeding population of birds, particularly in those species, like the common kestrel, that do not build their own nests. Kestrels actually use to breed in old corvid nests or in holes in buildings. By provisioning kestrels with artificial nest boxes, it is possible to increase the number of breeding individuals and, possibly, the population size. However, a number of factors need careful consideration to evaluate *a priori* the characteristics of nest boxes and where to install them and to assess *a posteriori* the effects of the nest box provisioning on the reproductive ecology and population dynamics of kestrels.

4.2 Breeding Density

Estimating breeding density is an important task in most population studies of birds. The breeding density indicates the number of active breeders in a given area. However, this number does not indicate the population size, because in the area there might also be individuals that did not mate or that mated but did not lay. This also means that only counting individuals rather than also counting active pairs would be misleading. The size of the area and the spacing of nests are also two major factors to assess carefully because they might strongly bias estimates of breeding density and make comparisons among studies difficult to achieve. Village (1984, 1990) estimated that a study area ranging from 80 to 200 km^2 might be suitable to make robust density estimates because he found low correlation between density and study area within that size range.

Density estimates of breeding kestrels vary significantly across localities (e.g. from 12 to 32 pairs per 100 km^2 in Village, 1990; 12 pairs per 100 km^2 in Brichetti & Fracasso, 2003; 19 pairs per 100 km^2 in Durany et al., 2003; 70 pairs per 100 km^2 in Mikeš, 2003; from 1.4 to 200 pairs per 100 km^2 in Thiollay & Bretagnolle, 2004; 120 pairs per 100 km^2 in Riegert et al., 2007a; from 88.5 to 122.2 pairs per 100 km^2 in Sumasgutner et al., 2014b). The question, then, is to identify the factors that affect breeding numbers because this knowledge is central for understanding population dynamics. Availability of food, shortage of nesting sites, winter mortality, emigration, immigration, exposure to pesticides and both inter- and intraspecies competition are certainly important factors to consider (Chapters 2, 6, 8 and 10). Cavé (1967) found that breeding densities were lower when rainfall was abundant, possibly because of a detrimental effect of weather on food supply. Village (1990) recorded the highest densities of breeding kestrels in habitats where prey were more abundant. Korpimäki and Norrdahl (1991) found that annual changes in the number of breeding birds were synchronous with those in the estimated density of prey. Also, breeding densities were larger in urban sites than in rural sites (e.g. Sumasgutner et al., 2014b) and were negatively influenced by the presence of large forests (e.g. Rodríguez et al., 2018).

Shortage of nesting sites is one major limitation of both breeding density and population stability of kestrels, as well as of many other birds of prey. Environments often have the topographical and trophic characteristics that could support a large breeding population, but a shortage of suitable sites for nesting, particularly in landscapes heavily modified by human activities, strongly limits breeding opportunities (Newton, 1979; Village, 1990).

4.3   Nesting Sites

Kestrels do not build their own nests. They rely instead on a wide range of nesting sites, including the disused nests of other bird species (e.g. corvids), tree holes, cavities in buildings or ledges on cliffs (Figure 4.1). Sometimes, kestrels may also breed on the ground if there are no suitable nesting sites or ground predators. This is the case on the Orkneys, where kestrels have been observed to breed in long tunnels in heather or rabbit burrows since 1945 (Balfour, 1955). Thus, kestrels are very versatile when it comes to breeding, which might be one reason for their ability to dwell in different habitats.

In Great Britain, Village (1983, 1990) found that kestrels used to breed in the abandoned nests of several bird species, such as the carrion crow, the magpie, the rook, the sparrowhawk, the long-eared owl, the pigeon and the heron. Tree holes are also commonly used. Village (1990) observed that the frequency with which particular tree species were used depended on their local availability and their propensity to form suitable holes. In English farmlands, Village (1990) found that ash trees provided 62% of holes for kestrels, while elm, oak and black poplar provided 25%, 3% and 3% of suitable holes, respectively.

Although common kestrels seem to use any kind of suitable nesting site available, some observations indicated that kestrels might also have some preferences. Village (1990) observed that in some areas the most common suitable nesting sites were not used as expected. If there is actually a preference, the question then is whether such preference has a genetic component or if it depends on early life experience, i.e. on the kind of nest the kestrel was reared in and learned to recognise. Experimental work on American kestrels did not find

Figure 4.1 Examples of natural nests: (a) cavity in a building; (b) abandoned stick nest built by hooded crow (*Corvus cornix*) on a pylon of power lines. Photographs by David Costantini.

evidence for nest-site preferences (Shutt & Bird, 1985), but American kestrels almost exclusively nest in holes. Another important question is whether there is variation among nesting sites in reproductive traits. Some studies found low variation in breeding parameters or fledging success across different types of nests (Riddle, 1979; Village, 1990; Krueger, 1998). Other studies found significant differences among nest types. In common kestrels from Finland, Germany, Isral and Spain, breeding success was higher in closed-type nests (nest boxes or cavities in buildings or walls) than in open-type nests (pre-existing stick nests), presumably because open nests are more vulnerable to predation, extreme weather conditions, or other environmental factors (e.g. Kostrzewa & Kostrzewa, 1997; Fargallo et al., 2001; Charter et al., 2007b).

4.4   Nest Availability and Nest Boxes

Village (1990) recognised three main dispersion patterns of nests. The first pattern is characterised by *irregular spacing*, where distance between neighbouring pairs varies and is usually more than a few hundred metres. This pattern is strongly dependent on the irregular distribution of available nesting sites. In farmlands, Village (1990) found that the distance between nests varied from 40 m to over 5 km, with an average of about 1–1.5 km and less than 7% of pairs nesting within 200 m from each other. The second pattern is *clumping breeding*. This occurs where nesting sites are close to each other and a large number of pairs is forced to breed less than 200 m from each other (Village, 1990). This pattern appears more common in suburban or urban areas (Tinbergen, 1940; Parr, 1969; Kurth, 1970; Riddle, 1979). Finally, *colonial breeding*, although rare, has been recorded in several countries. This pattern describes those situations where many pairs nest a few metres away from each other. For example, breeding colonies of common kestrels were found on rock faces, rookeries, sandstone quarries, gasometers or bridges (Fennel, 1954; Hagen, 1952; Hyuga, 1956; Peter & Zaumseil, 1982; Piechocki, 1982; Bustamante, 1994). The reasons for the colonial breeding are currently unknown. It might be that the strategy of colonial breeding is advantageous in places where nesting sites occur in limited supply and food is abundant. A large availability of prey would relax food competition, thus the aggressive behaviour of a pair against neighbours would be reduced as a consequence. It is clear that the availability of nesting sites and their relative position in the landscape have a significant effect on the reproductive behaviour of kestrels.

The use of artificial nest boxes has proved to be an effective tool in limiting the negative effects on reproduction caused by human-induced changes in land use/cover and the consequent decrease in nesting sites. The use of artificial nest boxes has also led to significant progress in bird conservation and in our understanding of the functional and evolutionary ecology of free-ranging birds of prey that exploit cavities for roosting and reproduction (Lambrechts et al., 2012). Nest boxes and their improved accessibility have also made easier to perform comparative and experimental field investigations.

Nest boxes are artificial cavities that have been designed to attract secondary cavity nesters for breeding (Figure 4.2). The widespread use of nest boxes can halt population declines or can considerably increase the size of a population, especially in environments where cavity-forming trees are missing or abandoned buildings have become unavailable (Lambrechts et al., 2012). There are several criteria for selecting appropriate sites to install nest boxes that need careful consideration; for example, habitat characteristics around the potential site, abundance of prey, size or orientation of the nest box. The *a posteriori* assessment of the success of nest boxes in restoring the local population also needs to consider several factors carefully. For example, an increase in local breeding density after nest box provisioning might result from birds incoming from adjacent areas, where density would consequently decrease.

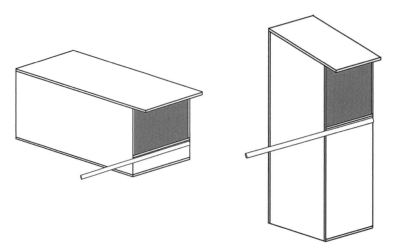

Figure 4.2 Examples of two different nest box designs for common kestrels. The size of the different components varies slightly across studies; we refer the readers to the literature presented in this chapter and to the caption of Figure 4.3.

Concerns about the generality and applicability of studies involving birds breeding in nest boxes have been raised because (i) the birds that breed in nest boxes may differ from conspecifics occupying other nest types and (ii) the characteristics of nest boxes may not mimic closely those of natural or preferred nesting sites (e.g. Møller, 1994; Lambrechts et al., 2012). The way nest boxes are designed, positioned, monitored and maintained may influence a cocktail of biotic and abiotic factors in the nest box chamber, probably also interacting with external environmental factors, such as food abundance or weather. Eventually, this might affect both reproductive decisions and success. Thus, the validity of ecological and evolutionary conclusions from data gathered using nest boxes might be questionable if comparisons with data collected from birds breeding in natural sites are not carried out.

Valkama and Korpimäki (1999) found that neither the orientation nor the size of the nest box affected important breeding traits, such as clutch size or number of fledglings, in common kestrels. The occupancy rate did not differ among nest boxes of different size (box volume: 17.2, 23.8 and 45.2 dm^3), but was lower in exposed than in sheltered nest boxes. Occupied nest boxes were further from forest edges, roads and houses and had more grassy ditches (where voles were more abundant) in their vicinity than unoccupied nest boxes. Also, the nest box size was not associated with body size or body condition of the parents. Although kestrels preferred to breed in sheltered nest boxes, there was not obvious benefit in terms of increased breeding success (Valkama & Korpimäki, 1999).

Fargallo et al. (2001) found that nest predation was higher in natural nests than in nest boxes. Fargallo et al. (2001) also observed several cases of starved nestlings within holes in buildings or of premature nestlings on the ground below nests located in holes. The mortality of nestlings was actually higher in nests located in holes of buildings than in nest boxes. Thus, the basal area of the nest and the depth of the hole might be very relevant drivers of breeding success. In Vienna, Kreiderits et al. (2016) found that closed nest structures had larger mean clutches than open nests (5.0 eggs for building cavities; 5.0 for nest boxes; 3.0 for crow nests, but 5.3 for open planters) and that the mean number of fledged offspring was greater for broods in nest boxes (4.0), planters (3.9) and building cavities (3.0), whereas the smallest mean was found in crow nests (0.8). Also, a recent review of 25 studies across 35 populations of common kestrel found that the average number of fledglings per reproduction was significantly greater in nest boxes (3.9) than in natural cavities (3.1) or open nests (2.5) (Fay et al., 2019). Such differences were owing to higher hatching success and nestling survival in nest boxes rather than an increase in clutch size (Fay et al., 2019).

The size of the nest box may also affect the degree to which kestrels compete with other species to acquire or defend it. For example, barn owls may evict kestrels that occupy the large nest boxes designed for them (Charter et al., 2007b, 2010). Similarly, kestrels may suffer competition from tawny owls in the occupancy of nest boxes (Dell'Omo et al., 2005). Also, the management of nest boxes during the non-breeding season has to be considered carefully. Sumasgutner et al. (2014b) showed that cleaning nest boxes before the start of the breeding season reduced the ectoparasite *Carnus hemapterus* burden of nestlings.

4.5   Power Lines as Breeding Sites

The scientific literature and newspapers have often dealt with the relationships between overhead power lines and the environment, mainly because of the potential risk of electrocution and collision that power lines represent for birds (Chapter 10). However, some features make overhead power lines attractive as nesting sites. Many reports showed that common kestrels and other kestrel species use to breed in abandoned corvid stick nests located on the pylons of power lines (see table 1 in Krueger, 1998).

Several authors suggested that the use of pylons of overhead power lines as a support for artificial nest boxes would provide suitable nesting sites for birds of prey (e.g. Goodland, 1973; Olendorff & Stoddart, 1974). In central Italy, a long-term project on the reproductive biology and ecology of common kestrels showed that pylons of overhead power lines can be successfully exploited to support nest boxes and establish a breeding population of kestrels (Dell'Omo et al., 2005).

In 1998, we installed 200 nest boxes on metal pylons of the ENEL/TERNA electricity network (Figure 4.3), with a voltage between 60 and 380 kV. Thirty-six different power lines were included in the work, crossing an area of about 1200 km^2. Nest boxes were made of marine plywood (thickness 1 cm; the size is reported in Figure 4.2), were open on the short side (similar to the model by Dewar & Shawyer, 1996), had drainage holes in the bottom and *Sphagnum* peat as litter. Nest boxes were attached to the pylons at various heights above the ground

Figure 4.3 Pylons of overhead power lines can be successfully exploited to support nest boxes and establish a breeding population of kestrel. The two images show examples of nest boxes used in a long-term project in central Italy. Nest boxes were made of marine plywood with a size of 50 × 30 × 30 cm (L × W × H). The roof was 60 cm long. Photographs by Gianluca Damiani (a) and David Costantini (b).

(4–23 m) and at various orientations in order to assess which factors might influence occupancy rate and breeding success.

We chose the pylons on which to attach the nest boxes following two strategies: (i) to distribute nest boxes as uniformly as possible around the city of Rome; (ii) to place a similar and large number of nest boxes across different environmental types (e.g. pasturelands, set-asides). Twenty of the 200 nest boxes were occupied by other species, namely house sparrow ($n = 11$), tawny owl ($n = 7$), Eurasian jay (*Garrulus glandarius*; $n = 1$) and western jackdaw ($n = 1$). Of the remaining 180 nest boxes, 52.6% was occupied in the first year after their installation. Nest boxes mounted at a height below 10 m above the ground were slightly less occupied than those placed between 10 and 15 m or above 15 m above the ground (Figure 4.4). Nest boxes oriented towards the north were less occupied, while those oriented with a southern exposure had the highest percentage of occupancy (Figure 4.5). Other nest boxes with different orientations had occupancy rates of about 50%. In addition, nest boxes located near forest patches or urbanised areas were the least occupied as previously found in Finland by Valkama and Korpimäki (1999).

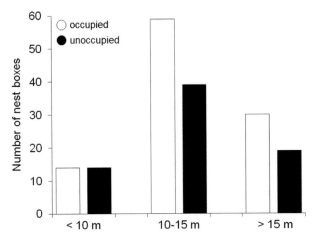

Figure 4.4 Percentage of occupancy of nest boxes by common kestrels in relation to height above the ground. Reproduced from Dell'Omo et al. (2005).

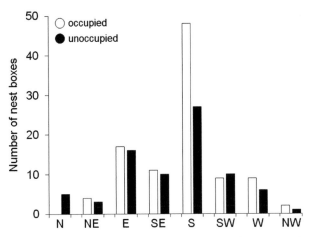

Figure 4.5 Percentage of occupancy of nest boxes by common kestrels in relation to orientation of entrance. Reproduced from Dell'Omo et al. (2005).

4.6   Conclusions

Breeding density is one important feature to quantify in order to better understand the population dynamics and to assess the conservation status of a species. There is a paucity of studies that quantified the breeding density of kestrels and identified the factors that may affect it. The availability of sites suitable for nesting is one major factor that may limit the size of the breeding population of kestrels. This is because kestrels do not build their own nest. Provisioning of artificial nest boxes has proved a valuable strategy to overcome the issue of nest availability. Publications and unpublished observations report that aspects of nest box design (e.g. size of the whole box, internal size of the nest cavity, presence or absence of drainage holes, wall thickness, type of bedding), location (e.g. nest box height above the ground, orientation of entrance hole, substrate to which the box is attached) and/or maintenance procedures (e.g. cleaning) may influence both the probability that a box is occupied and the expression of life-history traits of breeders (e.g. clutch size, hatching success, fledging success). However, these data are not always reported, which makes comparison of studies complicated. Before 1995, 61.5% of articles did not report any nest box characteristics, and this percentage increased to 73.8% in the period 1995–2012 (Lambrechts et al., 2012). We reiterate here the importance of indicating as many nest box characteristics as possible in order to enable to (i) properly assess if the nest box design affected the results and (ii) perform comparative studies while controlling for potential confounding effects due to variation among studies in nest box design.

5 Colourations, Sexual Selection and Mating Behaviour

5.1 Chapter Summary

Production of secondary sexual traits is a key component of reproductive investment in many sexually reproducing species. In this chapter, we have illustrated the secondary sexual traits and behaviours that are implicated in mate choice; the potential meaning of colourations of females and of their eggs as postmating sexual traits; and the signalling role of colourations in young kestrels. We have also described the biochemical bases of body colourations and the physiological costs associated with their production.

5.2 Introduction

In observing the marked differences in shape, size and colour between males and females that characterise many animal species, Darwin (1871) developed the idea that they might have been caused by competition between conspecific individuals to acquire a mate. Darwin realised that this competition was an opportunity for selection to drive the evolution of sexual traits that are important in mate choice. The production of secondary sexual traits is a key component of the reproductive investment of males in many sexually reproducing species (Darwin, 1871; Andersson, 1994). Secondary sexual traits can occur in a large variety of forms, such as visual (e.g. body colourations) and acoustic (e.g. song). Several sexual characters occur simultaneously, meaning that selection might operate on a package of multiple characters. Several conflicting hypotheses have been proposed to explain the adaptive value of displaying multiple sexual traits and how these traits are maintained by sexual selection (Zuk et al., 1992; Møller & Pomiankoski, 1993; Johnstone, 1996; Candolin, 2003; Lozano, 2009). The *multiple messages hypothesis* states that each secondary sexual trait signals a different quality of the male. The *redundant signal hypothesis* states that each trait provides information about the overall individual quality, thus errors in assessment of signals by the female may be reduced. The *unreliable signal hypothesis* states that a female may need to look at multiple traits because one trait could be an unreliable indicator of overall male quality. Finally, the *sexual interference hypothesis* proposes that additional male signals evolve as a way for males to hinder female mate choice by interfering with the propagation and reception of other males' sexual signals. On the other hand, females would have responded by evolving the ability to glean meaningful information from male signals.

A pillar of all these hypotheses is that a secondary sexual character signals the state or a given quality of the male. Thus, variation among males in the conspicuousness of ornaments would reflect that in individual phenotypic or genetic qualities. In turn, females would use this information to choose a high-quality mate that will bring to her and her offspring some fitness benefits. On the other hand, it may be more important to recognise low-quality males in order to not mate with them; in so doing, there will be a reduced likelihood of large penalties in fitness (Gomes & Cardoso, 2018).

The fitness benefits of having a high-quality partner might occur as contribution to parental care, provision of a good rearing environment, genetic benefits to the off-spring, or avoidance of infectious diseases. It follows that sexual signals of males have to be honest, that is the information they convey about the male's state needs to be reliable, otherwise the ornament would be quickly counter-selected if males were attempting to cheat (Zahavi, 1975; Zahavi & Zahavi, 1997).

5.3   Sexual Dimorphism and Feminised Phenotypes

Common kestrels show a marked sexual dimorphism in both body size and colourations (Figure 5.1). Males are around 20% smaller than females (Village, 1990), but large variation occurs within and among populations (Table 5.1 and Figure 5.1). Adult males are brownish-red to brick-red with black spots on their backs and on the upper sides of their wings. Females have brown backs and upper sides of their wings. Males have completely brown or blue–grey heads, while females have brown heads. Males have rumps, upper tail coverts and tail feathers from brown–grey to completely grey and mainly unbarred, while females have black-barred rumps and tails that vary in colour from brown to grey. The tail has also a black tip with a narrow white rim in both sexes. Melanins are the pigments responsible for the feather colourations, while carotenoids are the pigments used to produce the colouration of the skin of feet, cere and eye ring.

Although juvenile males are more similar to adult females than to adult males, they still show slight sexual dimorphism in both body size and colourations (Village et al., 1980; Village, 1990). There are, however, large differences among male juveniles in the way they look. Some males are difficult to sex visually because they express feminised phenotypes in young age and have delayed plumage maturation (Village et al., 1980; Village, 1990; Hakkarainen et al., 1993). Why feminised phenotypes are maintained within kestrel populations remains an open question. Earlier work carried out in Finland showed that female-like young males bred closer to adult males than did other adult males and, in so doing, they had a higher probability of breeding success-fully (Hakkarainen et al., 1993). Moreover, in a mate choice experiment carried out in captivity, adult males could not discriminate between a female and a female-like young male (Hakkarainen et al., 1993). Thus, rather than delaying the sexual maturation as they did for the plumage, female-like young males could breed successfully in their

Table 5.1 Examples of studies on common kestrels in Europe that illustrate the variation in body mass between males and females across seasons and geographic regions; it ranges from around 150 to 260 g for males and from 170 to over 300 g for females. n = sample size.

Stage of life	Mean female	Mean male	Min–max female	Min–max male	Female (n)	Male (n)	Region	Study
Mating	226	219	170–260	185–255	8	7	Italy, 44°48′N, 10°20′E	Costantini et al., 2014
Nestling rearing	223	174	195–270	155–205	19	15	Italy, 44°48′N, 10°20′E	Costantini et al., 2014
Non-breeding	244	203					The Netherlands, 53°20′N, 6°16′E	Masman et al., 1986
Mating	263	206					The Netherlands, 53°20′N, 6°16′E	Masman et al., 1986
Laying	305	213					The Netherlands, 53°20′N, 6°16′E	Masman et al., 1986
Incubation	275	204					The Netherlands, 53°20′N, 6°16′E	Masman et al., 1986
Early nestling rearing	267	188					The Netherlands, 53°20′N, 6°16′E	Masman et al., 1986
Late nestling rearing	235	196					The Netherlands, 53°20′N, 6°16′E	Masman et al., 1986
Post fledging period	197	173					The Netherlands, 53°20′N, 6°16′E	Masman et al., 1986
Post reproductive moult	229	202					The Netherlands, 53°20′N, 6°16′E	Masman et al., 1986
Non-breeding	255, 227	213, 200			8, 33	10, 45	Scotland, England	Village, 1990
Mating	220, 239	208, 197			3, 31	17, 55	Scotland, England	Village, 1990
Laying	284, 264	208, 193			9, 15	7, 10	Scotland, England	Village, 1990
Incubation	276, 266	209, 192			78, 96	31, 22	Scotland, England	Village, 1990
Early nestling rearing	246, 251	198, 192			7, 24	7, 17	Scotland, England	Village, 1990
Late nestling rearing	246, 233	201, 180			19, 11	16, 10	Scotland, England	Village, 1990
Nestling rearing	215	180	186–258	150–218	133	113	Finland, 62°59′–63°10′N, 22°50′–23°20′E	Jönsson et al., 1999

Figure 5.1 Kestrels are medium-size raptors approximately 32–38 cm long, with a wingspan of around 68–78 cm. Females are approximately 20% larger than males, but overlap occurs between the smaller females and the larger males. This image illustrates a large within-sex variation in body size (the first three kestrels from the left are males). Specimens are from the collection of the Muséum National d'Histoire Naturelle (MNHN; Paris, France). The MNHN gives access to the collections in the framework of the RECOLNAT National Research Infrastructure. Photographs by David Costantini. A black and white version of this figure will appear in some formats. For the colour version, refer to the plate section.

first year of life; this strategy could be favoured by a reduced agonistic behaviour with adult males, which do not recognise them as competitors. It might be that female-like males either avoided the costs of agonistic behaviour with older males or were less exposed to predation.

Vergara and Fargallo (2007) analysed aggressive and courtship behaviour of breeding males and females in the presence of an adult male, an adult female and a one-year-old either unmoulted (grey-barred rump) or moulted (plain grey rump) natural decoy. Male kestrels attacked adult male decoys more frequently than female and both moulted and unmoulted one-year-old male decoys and no sexual displays were observed. Females were more aggressive towards female and unmoulted one-year-old male decoys as compared to adult or moulted one-year-old male decoys (Figure 5.2).

Vergara and Fargallo (2007) also observed that females solicited copulas from moulted but not from unmoulted one-year-old male decoys. These results suggested that moulted young males could attract females and diminish the aggressiveness from males by honestly signalling their subordination or low competitive capacity. On the other hand, unmoulted young males, by resembling females, could obtain benefits during the winter, when their cryptic plumage would reduce the probability of being attacked by adult males (Vergara & Fargallo, 2007).

Further work carried out in Spain found that conspecific aggressiveness of both males and females might affect the timing of plumage maturation of males (Vergara et al., 2013). It has also been found that the more feminised male juveniles had higher chances of survival, and thus higher chances to recruit into the population, than the less-feminised male juveniles (López-Rull et al., 2016). Higher parental care for the more feminised males might explain their higher survival probability; however, this does not seem to be supported by empirical data (e.g. Laaksonen et al., 2004b; Vergara & Fargallo, 2008b). Another reason for the higher survival of more feminised males might lie with the costs of developing a male phenotype that are paid later in life. There is, however, little information on winter biology, dominance hierarchies and predation risk of kestrels. Moreover, in those geographic areas where kestrels migrate, relying on two different migration phenologies might reduce competition between adult and young males. For example, Village (1985a) observed that resident first-year males waited longer than adults before finding a mate, probably because of the competition. On the other hand, migrating first-year males arrived at the breeding grounds later than adults, probably in order to avoid the peak of competition with adults. Well-designed experiments and long-term longitudinal data are clearly needed in order to elucidate the benefits of and mechanisms underlying the delay in the plumage maturity of males. This information is important to understand how different male phenotypes and reproductive tactics are generated and maintained within populations.

5.4   How Do Female Kestrels Choose Their Mates?

Kestrels are monogamous, but cases of males mated contemporarily with two females breeding in two separate nests (MacDonald, 1973; Korpimäki, 1988; Village, 1990; Wang et al., 2019) or raising the young together in the same nest (Motti et al., 2008)

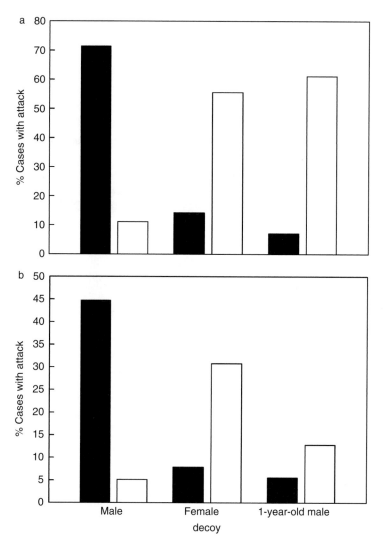

Figure 5.2 Percentage of cases with attacks towards each decoy for breeder male and breeder female kestrels in the (a) unmoulted (i.e. a one-year-old male with a grey-barred rump) and (b) moulted (i.e. one-year-old male with a plain-grey rump) treatment. Reprinted from Vergara and Fargallo (2007) with permission from Elsevier.

were reported in the literature. There is a clear division of labour between mates. The male provides food to her mate from before the start of egg laying until the mid-stage of the nestling-rearing period. The male is also the main food provider for the offspring until they are about 2 weeks old. Females incubate, take care of the offspring and hunt mostly in the second part of the nestling period. Thus, it would be expected that females choose those males that are better at catching prey and providing food.

Calls and territorial displays are two important aspects of mating and courtship periods (Tinbergen, 1940; Village, 1990), but they are not directly linked to food

Figure 5.3 A male is passing a prey to his mate. In kestrels, the male provides food to the female from before the start of egg laying until nestlings are in the middle of their growing period. Photograph by Gianluca Damiani. A black and white version of this figure will appear in some formats. For the colour version, refer to the plate section.

availability. Male kestrels also provide food to females during the courtship period (*prey courtship feeding*). The male arrives at a perch near the nest and, straightaway, calls the female, who flies towards the male to collect the prey (Figure 5.3). Also, the male can fly directly to the nest, where it passes the prey to the female. If food pass indicates male hunting skills, it would be expected that females choose those males that are good food-providers. However, Palokangas et al. (1992) did not find any evidence that female kestrels choose males that are good in providing food. In contrast, the food provisioning behaviour might have been selected because it improves the body condition of females before the laying and incubation periods. Thus, food pass from the male to the female would not be used by the female to make a choice, but it might be that the female's mate choice relies on traits that are indirectly linked to male hunting or parental skills. The question then is which other traits a female uses to choose her mate.

Early work on the mate choice in the common kestrel carried out in Finland showed that female's choice varied under different availability of mating options (Palokangas et al., 1992). The importance of a given trait value in affecting mate choice might, therefore, vary among years depending on the variation among males in the expression of that particular trait. Palokangas et al. (1992) observed that during those years of large availability of voles there were many more potential males for each female to choose and that in those years males with long tails were more likely to be chosen than males with short tails. However, in the poor vole year, there was no evidence for a female's choice of males with longer tails. These results suggested that females may be choosy with respect to male traits only if the density of male kestrels is high, providing the female with alternative mating options. In the year when females appeared to be choosy for long-tailed males, the density of unmated males for females was, on average, 1 per 10 km^2, while in the poor vole year such a density was just 0.2 males per 10 km^2 (Palokangas et al., 1992). These data did not prove, however, that females were actually using the tail length as a clue for mate choice because tail length might have covaried with other important male attributes, such as display performance, body condition or body colourations. Moreover, tail length was not correlated with feeding effort or with breeding success, indicating that the female's choice for long-tailed males might not have brought any clear direct fitness benefits (Palokangas et al., 1992).

Later work suggested that female kestrels might gain fitness benefits by choosing brightly ornamented males (Palokangas et al., 1994). Work carried out on free-ranging kestrels in Finland showed that males with a higher brightness index (combining the back and tail colours and the size of back-spots) brought more prey to the nest and had more offspring (Palokangas et al., 1994). Importantly, the differences in reproductive success between brighter and paler males were not due to their mates investing differently into reproduction because they invested similarly. The higher reproductive success of brighter males was also confirmed by another study on Finnish kestrels; however, brighter males did not invest more in hunting effort, nor did they deliver more prey than dull males did (Tolonen & Korpimäki, 1994). This result would suggest that brighter males might provide some fitness benefits that are not linked to food provisioning.

In a mate-choice experiment carried out in captivity, female kestrels preferred to approach brighter males (Palokangas et al., 1994). Although there were some discrepancies between data collected in the field and in the captivity experiment (possibly due to restricted variation in brightness index in captive males or different light conditions that affected the brightness itself), these data suggested that female kestrels pairing with brightly coloured males may gain reproductive benefits. It is, however, unclear why males with blue–grey tail feathers had more *Haemoproteus* blood parasites than those with brown-tailed feathers (Korpimäki et al., 1995). If brighter males are of higher quality and more likely to be chosen, brighter males should have fewer parasites than paler males according to the *Hamilton–Zuk hypothesis* (Hamilton & Zuk, 1982). It might also be that brighter males may tolerate larger parasite loads than duller males and the cost of producing bright colours. However, females mated with infected males delayed the start of egg laying and laid a lower number of eggs than those mated with non-infected males (Korpimäki et al., 1995). Thus, the choice of mating with brighter males might have been maladaptive (Korpimäki et al., 1995). The lack of data on other parasites and of experimental manipulations limit the inferences about the implications of these results. Moreover, the colour brightness was estimated visually; this method might not be adequate given that the vision of kestrels differs from that of humans.

In the American kestrel, no male trait was clearly related to the time that males remained unpaired after arriving on their breeding territories (*mating lag*), possibly because of the scarcity of unpaired males, and the fact that there was very little variation in mating lags (most pairs were formed within one day of the male arrival; Wiehn, 1997). However, females mated to males with either a bright plumage or a narrow subterminal tail band had higher reproductive success (Wiehn, 1997). Previous work on American kestrels found that mating was assortative by body condition in free-living individuals (Bortolotti & Iko, 1992), whereas a study on captive kestrels indicated the importance of frequent nest-site inspections by males in mate choice (Duncan & Bird, 1989).

These experiments did not provide a definite experimental demonstration of whether the overall brightness of the plumage colouration or only single plumage components (e.g. grey intensity, size of black spots) plays any major role in mate choice. To fill in some of these gaps, Zampiga et al. (2008a) enhanced experimentally the grey colouration of both tail and head of males and tested whether males with enhanced colouration were more likely to be chosen than those with unchanged colouration. The mate choice trials showed that females did not have any preference for the males with enhanced colouration; rather, female kestrels were constantly attracted to one particular male irrespective of its grey colouration (Zampiga et al., 2008a). Although the sample size was small, these results suggested that other traits might be involved in mate choice. For example, the visible skin parts of common kestrels are yellow because of the deposition of carotenoids in skin cells, and significant variation in the intensity of these colourations occurs among males (Section 5.5).

Another important male attribute that might be involved in mate choice is body size, because it is a trait that can be easily evaluated by the female in a short time. Both energy expenditure and molecular oxidative damage (Chapter 7) are much greater in male than in female kestrels during the nestling-rearing period (Masman et al., 1989; Casagrande et al., 2011). If smaller males face lower hunting physiological costs and are, consequently, more efficient in hunting, this might bring some benefits to the female. In a mate choice experiment carried out on captive kestrels, Hakkarainen et al. (1996) found that females actually preferred smaller males, which also had better hunting success under captivity conditions. Moreover, Hakkarainen et al. (1996) observed that free-living smaller males provided females more food than larger males during the courtship feeding period in poor (when selection on hunting efficiency of males is expected to be high) but not in good vole years.

It is important to consider that most knowledge about mate choice in kestrels is based on populations living at high latitudes and this might not apply to populations at lower latitudes. This is because northern kestrels migrate and mate choice is made within a few days after both males and females are back at the breeding grounds (e.g. Palokangas et al., 1992). Thus, females have to screen a male and make a choice quickly because they are on quite a strict schedule. In southern Europe, many kestrels stay on the breeding grounds all year round, and the duration of the breeding season may be longer compared to that of northern populations. Thus, southern female kestrels might have more time available to make a choice due to being on a less-strict temporal schedule compared to northern kestrels. Mate choice tactics might therefore vary latitudinally.

5.5   Carotenoid-based Colourations

5.5.1   Carotenoids in the Kestrel

Carotenoids are lipophilic pigments that animals are unable to synthesise; hence, they must be acquired from food (Hill & McGraw, 2006a, 2006b). Carotenoids form the basis of many yellow, orange and red body colourations in the animal kingdom, including birds (Hill & McGraw, 2006a, 2006b). In birds, carotenoid pigmentation is more common in the bare parts than in the plumage, and yellow colourations appear to be more widespread than red ones (Olson & Owens, 2005).

Given the limited availability of carotenoids in the food and their multiple physiological functions (e.g. immune stimulation, antioxidant protection), a large allocation of carotenoids to the production of sexual colourations would mean that less is available to be allocated to other functions (Lozano, 1994; Møller et al., 2000; Hill & McGraw, 2006a, 2006b). Thus, a high allocation of carotenoids to body colourations might come at a cost (e.g. a less-efficient immune system) for low-quality males or for all males that only high-quality males might be able to sustain.

All kestrel species display carotenoid-based colourations in the skin of lores, cere and tarsi. Laboratory analyses carried out using high-performance liquid chromatography showed that lutein was the predominant carotenoid (~90% of the total amount of carotenoids) in both the blood and the bare parts of common kestrels from a population of northern Italy (Casagrande et al., 2006b). Zeaxanthin was the only additional carotenoid detected in both blood and skin samples. In contrast, the chromatographic analyses of the reddish back plumage did not reveal the presence of any carotenoids (Casagrande et al., 2006b). Another study carried out on kestrels in Spain also found that lutein was the dominant carotenoid in the blood (De Neve et al., 2008). However, they also found small proportions of β-carotene, cis-lutein and β-criptoxanthin, while zeaxanthin was not detected (De Neve et al., 2008). In agreement with the work on common kestrels, lutein was also the predominant carotenoid detected in the American kestrel (Bortolotti et al., 2000). Lutein and zeaxanthin were also found in the peregrine falcon, together with 3′-dehydrolutein, a carotenoid metabolically derived from the enzymatic dehydrogenation of dietary carotenoids (Costantini et al., 2009a). It is unclear why different carotenoid profiles were detected in two different populations of common kestrel. Further work will be needed to elucidate whether the different carotenoid profiles reported in Casagrande et al. (2006b) and in De Neve et al. (2008) were due to methodological reasons or to real differences between populations in the way carotenoids are absorbed from food and processed metabolically.

5.5.2   Colourations as Honest Sexual Signals

Casagrande et al. (2006b) found that male common kestrels had a brighter skin colouration than females, and that brighter males delivered a larger number of prey to the nest and held breeding grounds of higher quality in terms of prey availability. Although there is not yet any experimental proof that female kestrels choose males with brighter carotenoid-based colourations, Casagrande et al. (2006b) provided indirect evidence that the brightness of skin colourations could be positively associated with the foraging capacity of the male. This is very important in the kestrel because males are responsible for provisioning the female with food during the incubation phase, and both the female and offspring for around the first two weeks after egg hatching. Thus, choosing a male with good foraging skills or a high-quality home range in terms of prey availability would carry benefits for the entire family.

Brighter skin colouration of males compared to females was also found in other populations of common kestrel (Casagrande et al., 2011) or of American kestrel (Negro et al., 1998). However, such differences between sexes were apparent at the time of pairing, while they decreased abruptly along the breeding season (Negro et al., 1998; Laaksonen et al., 2008; Casagrande et al., 2011).

Carotenoids have several physiological functions. The question then is whether carotenoid-based colourations also signal some attributes of the male's physiological state. For example, it has been suggested that carotenoid-based colourations convey information about the male's resistance against molecular oxidative damage (e.g. that caused by free radicals on lipids, proteins and DNA) because carotenoids have anti-oxidant properties (von Schantz et al., 1999). In the common kestrel, neither correlative studies nor experimental administration of carotenoids revealed a significant antioxidant role of carotenoids in nestlings, which face the high metabolically demanding phases of growth and development (Costantini et al., 2006, 2007b). The oral administration of about 4 mg of carotenoids (lutein and zeaxanthin) every other day for a total of five administrations increased the serum concentration from an average of about 23 $\mu g\ ml^{-1}$ to an average of about 33 $\mu g\ ml^{-1}$ without causing any changes in a marker of serum oxidative damage or in a marker of serum non-enzymatic antioxidant capacity (Costantini et al., 2007b).

Other hypotheses have been proposed in order to explain why carotenoid-based colourations are common in birds and, seemingly, associated with mate quality and reproductive advantages. Hartley and Kennedy (2004) proposed that carotenoids are prone to oxidation and consequent bleaching, thus they might signal the availability of other, more potent and limiting antioxidant resources, rather than being involved in physiological trade-offs themselves (*protection hypothesis sensu* Pérez et al., 2008). Second, if carotenoids are wholly or partly the resource being advertised, then they are signalling other qualities, such as their contribution to cell signalling, gene activation, regulation of immune activity, tissue repair, or embryonic development (*alternative function hypothesis sensu* Metcalfe & Alonso-Alvarez, 2010). Irrespective of the actual mechanism, a brood size manipulation experiment showed that while an enlarged brood size increased the stress level of both parents, it did not affect the cere hue nor the plasma concentration of carotenoids of both males and females (Laaksonen et al., 2008). This result might suggest that kestrels actually protected carotenoid colourations from bleaching or that these colourations do not have any signalling role, at least during the nestling-rearing phase when signalling is less important than during the mating phase.

Another hypothesis proposes that carotenoids might exert detrimental effects on cell membranes, potentially acting as pro-oxidants (*toxicity hypothesis*) when occurring in the body at high concentrations (Zahavi & Zahavi, 1997; El Agamey et al., 2004; Costantini et al., 2007d) or under certain contexts (Simons, 2013; Beamonte-Barrientos et al., 2014). Consequently, individuals could use carotenoid-dependent traits to reveal their capacity to endure this handicap (Metcalfe & Alonso-Alvarez, 2010). This hypothesis is intriguing, but its ecological relevance is unclear. A study on captive common kestrels showed that the toxic effects of carotenoids might occur in this species (Costantini et al., 2007d). A prolonged carotenoid supplementation through the diet increased the blood oxidative damage and reduced the body condition. The serum concentration of carotenoids in supplemented birds (Chapter 7) was particularly high if compared to free-living birds (Casagrande et al., 2006b, 2007, 2011), making it unclear if such detrimental effects occur in free-living kestrels. Further work will be needed to assess the relevance of the toxicity hypothesis in kestrels.

5.5.3   Hormonal Regulation of Colourations

Developing a general explanation for the role of carotenoid-dependent colourations in mate choice requires a better understanding of the mechanisms that regulate their expression. Androgens (male steroid hormones) can affect the expression of skin carotenoid-based colourations in a number of bird species (e.g. Eens et al., 2000; Blas et al., 2006; McGraw et al., 2006; Mougeot et al., 2007). The skin colouration varies with sex, season and individual age in the American kestrel (Bortolotti et al., 1996; Negro et al., 1998) as do the androgen levels in many species with bi-parental care (Kimball, 2006). The skin colouration of American kestrels was not, however, associated with androgen levels (Bortolotti et al., 1996). One reason for this lack of correlation between carotenoids and androgens might lie with the effects that androgens may have on other functions, such as immunity and regulation of the oxidative status, which in turn are functionally linked to some degree to carotenoid metabolism (e.g. Folstad & Karter, 1992; Alonso-Alvarez et al., 2007). Casagrande et al. (2011) found that skin colouration was more intense in males than in females during the mating season, but not during the nestling-rearing phase, in both free-ranging and captive common kestrels. The plasma concentration of carotenoids was associated with the sexual and seasonal variation in skin colouration, while neither circulating levels of cholesterol nor oxidative status markers were associated with skin colouration. These results would undermine the hypotheses that both circulating lipids and oxidative stress may be limiting factors of carotenoid-dependent trait expression (Casagrande et al., 2011). Regarding hormones, changes in dihydrotestosterone (a hormone metabolically derived from testosterone), but not in testosterone or estradiol, were associated with those of skin colouration in captive male kestrels. Casagrande et al. (2011) also found a seasonal decline of skin colouration in male kestrels similar to that previously recorded in the American kestrel (Negro et al., 1998), even if the amount of carotenoids in the diet did not vary across the study period. These results would suggest that there might be an endogenous regulation of skin colouration irrespective of the availability of carotenoids in the food.

In a next study, Casagrande et al. (2012) experimentally manipulated the levels of either dihydrotestosterone (the androgenic pathway) or estradiol (the oestrogenic pathway) in order to test their effects on skin colouration, circulating carotenoids and plasma oxidative status in captive common kestrels. The experiment was carried out outside the breeding season in order to limit the interference of increased production of hormones and increased physical activity that both occur during the breeding season (Figure 5.4). Estradiol increased oxidative damage and reduced the non-enzymatic antioxidant capacity of plasma without affecting either skin colourations or circulating carotenoids in both males and females (Casagrande et al., 2012). In contrast, dihydrotestosterone did not affect the oxidative status or plasma carotenoids, but it increased the skin colouration in both males and females. Because testosterone can be converted to both estradiol and dihydrotestosterone, Casagrande et al. (2012) suggested that testosterone might be one mediator of the trade-off between the colour expression regulated by dihydrotestosterone and the oxidative stress induced by estradiol.

5.5.4   Genetic Basis of Colourations

In the context of sexual selection theory, it is very important to know the additive genetic component (i.e. narrow-sense heritability) of an ornament in order to estimate the indirect genetic benefits of mate choice. On the other hand, it is also important to know whether ornaments are shaped by environmental components, because this would provide clues about the direct non-genetic benefits of mate choice, such as parental care or resistance to parasites. Although the heritability of a secondary sexual trait is an important prerequisite to estimate its potential for evolution, in the case of low heritability there might still be the potential for evolution if the ornament were influenced by non-genetic inheritance (e.g. parental effects, epigenetics: Bondurianksy & Day, 2009; Danchin et al., 2011).

Casagrande et al. (2009) found that the environmental conditions experienced early in life explained the individual variation in the serum concentration of carotenoids and in a metric of skin colouration (hue) in nestling kestrels in central Italy. On the other hand, there was no evidence for a significant role of the genetic component. Similarly, Bortolotti et al. (2000) found that the variance in plasma concentration of carotenoids in nestling American kestrels was largely explained by the nest of rearing and not by the nest of origin, thus indicating a low genetic effect. However, the genetic component of variation in sexual traits can change with age (Robinson et al., 2008). Using a long-term data set based on adult individuals, Vergara et al. (2015) found that two

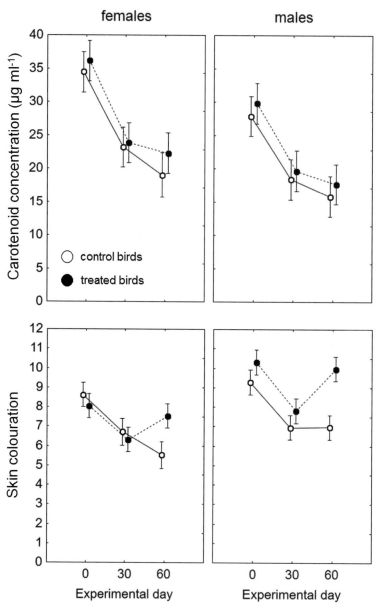

Figure 5.4 Plasma concentration of carotenoids and skin colour (red colorimetric component) before the start of the experiment (experimental day 0), after administration of 17β-estradiol started on day 0 (experimental day 30) and of 5α-dihydrotestosterone started on day 30 (experimental day 60). Values are shown as mean and standard error. Reprinted with slight modifications from Casagrande et al. (2012) with permission from Springer Nature.

colorimetric components of skin colourations, namely hue and brightness, showed very low levels of additive genetic variance, suggesting that the environment plays a major role in determining their expression. On the other hand, they found a significant moderate heritability of the chroma of the skin colouration. In addition, they found a significant negative directional selection on the hue and a positive directional selection on the chroma in females, indicating that natural selection might favour females with low hue and high chroma values (Vergara et al., 2015). Overall, these results highlighted the importance of measuring multiple metrics of colouration in order to get a better understanding of the evolutionary dynamics of carotenoid-based skin colourations. To this end, it will also be fundamental to know more about the exact biological meaning of each colour metric.

5.6   Melanin-based Colourations

Melanins are pigments that animals synthesise from the amino acids phenylalanine and tyrosine, which are obtained through the diet (Hearing, 1993). Melanins occur in two chemically distinct forms that generate different colourations. Eumelanin produces grey and black colourations, while pheomelanin generates reddish-brown and brown colourations. Melanic traits are under control of the pro-opiomelanocortin gene, which is responsible for the synthesis of melanocyte-stimulating hormones and the adrenocorticotropic hormone (Lin & Fisher, 2007). A study on common kestrels housed in captivity recorded a total melanin concentration in feathers of around $60 \ \mu g \ mg^{-1}$ and, as expected, confirmed that eumelanin and pheomelanin were the dominant forms in visibly dark and red regions of feathers, respectively (Edwards et al., 2016).

5.6.1   Role of Melanic Traits in Young and Adults

Several lines of evidence show that melanic traits might function as signals of phenotypic quality and might be affected by the environment in young common kestrels. Fargallo et al. (2007a) found that the proportion of greyest-rumped male nestlings in years of large vole abundance was higher than during years of low vole abundance and that the rearing environment had a significant effect on the expression of melanin-based grey colouration of male nestling kestrels. In another experiment, Fargallo et al. (2007b) found that experimentally increased levels of testosterone (i) negatively affected both the body condition index (a proxy of energy reserves) and the cell-mediated immune response and (ii) diminished the expression of grey rump coloration in male nestling kestrels. Parejo and Silva (2009b) found that nestlings supplemented with the amino acid methionine developed less-intense brown plumage on their backs compared with control nestlings, indicating that the quality of diet might significantly affect the expression of a melanic trait. It has also been found that juvenile melanin colourations were associated with several behavioural traits, such as dominance rank (Vergara & Fargallo, 2008b), anti-predator behaviour (van de Brink et al., 2012) and boldness (López-Idiáquez et al., 2019). Thus, juvenile melanin colourations might work as a badge of state in the context of sibling competition or parent–offspring communication.

In male adult kestrels, Parejo et al. (2011) found that several metrics of melanic colourations of head, back and rump were not associated with the body condition index, *Haemoproteus* infection or reproductive success. However, higher levels of circulating natural antibodies were associated with brighter and purer ultraviolet grey in the rump. Importantly, there was no evidence for selective mating regarding plumage colour metrics, indicating that the colour metrics measured might not have played any role in mate choice (Parejo et al., 2011).

5.6.2   Costs of Melanin Production

How the expression of melanin-based colourations is linked to other endogenous costs in kestrels has not been explored. For example, melanogenesis is also regulated by the availability of thiol groups, such as free cysteine and cysteine-containing peptides, which are used to synthesize pheomelanin (Ozeki et al., 1997; Benathan et al., 1999; García-Borrón & Olivares Sánchez, 2011). Thiol molecules, like glutathione, also play an important role in the regulation of the cellular oxidative status and protection against oxidative damages. Thus, oxidative stress might play some role in mediating the expression of melanin-based colourations (reviewed in Costantini, 2014).

5.6.3   Genetic Component of Melanic Traits

Pedigree-based quantitative genetic analyses showed that melanin-based colourations have a large and significant additive genetic component in both adult and nestling common kestrels (Kim et al., 2013) similarly to that estimated in two nocturnal birds of prey (barn owl in Antoniazza et al., 2010; tawny owl in Karell et al., 2011). Annual variations in environmental conditions influenced to some degree the expression of colourations (Kim et al., 2013). Further analyses carried out on males showed (i) a negative genetic covariance between melanin-based colouration and body mass and (ii) increased reproductive fitness in adults with intermediate phenotypic values of melanin colouration and body mass, suggesting that stabilising selection might have affected the dynamics of melanin colourations in the population (Figure 5.5).

Cross-sectional (between-individual) and longitudinal (within-individual) analyses based on data collected for 10 years showed that (i) the number, but not the size, of black spots decreased with individual age in males, (ii) the size of spots was larger in good vole years, (iii) the reproductive performance of males increased with age when there was a shortage of food and (iv) females obtained reproductive benefits when mating with older and less-spotted males under low-food conditions (López-Idiáquez et al., 2016b). Overall, this study suggested that the number of spots might convey information about the genetic quality, while the size of spots might provide informa-tion about the capacity of the male to hold a high-quality territory (e.g. with a larger availability of prey). In other words, females would prefer males with few but large spots to those with many but small spots. This scenario would agree with the *multiple messages hypothesis*, whereby different aspects of a secondary sexual trait convey information about different qualities of the male. Moreover, the decline in the number

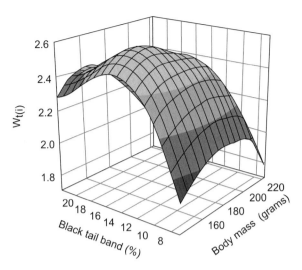

Figure 5.5 Males with intermediate values of the black tail band (% of its width in the whole tail length) and body mass had higher fitness ($W_{t(i)}$; see Brommer et al., 2007). Reprinted with slight modifications from Kim et al. (2013) with permission from Springer Nature.

of back spots in males was not apparently due to senescence because the reproductive output did not decline. However, reproductive senescence might become evident at older ages than those considered in the study.

5.7   Interaction between Carotenoid- and Melanin-based Traits in Males

We know very little about the extent to which different secondary sexual traits covary with each other and which information about the individual quality they convey. This information is fundamental to understand whether multiple ornaments are maintained and evolved and, if so, how and why. Common kestrels have both static and dynamic colourations. The colouration of the plumage is static because it does not change quickly. It is fixed at moulting and it fades in the long term. Thus, it should convey mainly information about the condition of the male during the moulting period. This period differs, however, among groups of feathers. For example, tail feathers are moulted after the reproductive period, while back and rump feathers can be moulted even during the last stage of the nestling-rearing period (Village, 1990). Thus, different feather groups might convey different information about the male's quality. Wear and tear of the feather might still be important to whether fading of the colouration is linked to the quality of the feather itself (e.g. amount of pigment deposited, keratin structure). However, we do not have any experimental investigation of this hypothesis.

On the other hand, the carotenoid-based skin colourations are very dynamic. For example, supplementation of carotenoids through the diet increased the colouration of tarsi within 7–14 days (Figure 5.6) from the start of the administration (Casagrande et al., 2007; Costantini et al., 2007d). Studies on northern populations of kestrels found

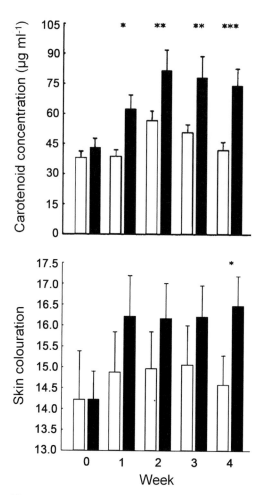

Figure 5.6 A daily supplementation of carotenoids significantly increased their plasma concentration in captive wild common kestrels as compared to control birds (i.e. those that were not supplemented with carotenoids). In contrast, the skin colouration (red colorimetric component) of supplemented birds was significantly higher than that of control birds after only 4 weeks of carotenoid supplementation. Values are shown as mean and standard error. Reprinted with slight modifications from Costantini et al. (2007d) with permission from Springer Nature.

that females might take a few days to choose their mate (e.g. Palokangas et al., 1992). This time window appears to be too short to detect a significant change of skin colouration due to the male being exposed to some kind of stressful condition at the breeding ground. However, any changes of skin colouration might reflect the costs that males experience while migrating.

There is only one study that has quantified multiple melanin- and carotenoid-based traits in the same individuals. This work suggested that different colourations might convey different information about the male's quality as proposed by the *multiple messages hypothesis* (Vergara & Fargallo, 2011). The colourations of bill cere (carotenoid-based) and of the rump (melanin-based) were both negatively and significantly correlated with the male body condition index. However, bill cere and rump colourations were not significantly correlated with each other. It has also been suggested that the rump colouration of females might be a badge of social status. However, the rump colouration of females was not correlated with fitness traits, such as clutch size or body condition index. Given these results and the moderate sample size, it is premature to infer about the roles that different colourations play in sexual or social signalling. We also lack information about the physiological mechanisms that underlie the expression of different colourations or the differential sensitivity to wear and tear that feathers might have.

5.8   Meaning of Female Colourations

5.8.1   Body Colourations

The origin and meaning of female colourations have received comparatively less attention than those of males (Cunningham & Birkhead, 1998; Amundsen, 2000; Kraaijeveld et al., 2007). Vergara et al. (2009) suggested that the colouration of the rump might be environmentally constrained because they found a positive covariation between the colouration of the rump and the environmental conditions (or reproductive investment) of the previous year quantified as clutch size at population level. Parejo et al. (2011) also found that specific metrics of colouration may provide complementary information about the female's qualities. They found that females with (i) higher scores of brightness for the head colouration and higher ultraviolet chroma scores of the black part of the rump were in better body condition and (ii) lower scores of brightness and higher ultraviolet chroma scores of the black part of the rump had higher circulating natural antibodies (Parejo et al., 2011). The link between some metrics of rump colouration and female condition might depend mainly on the current environmental conditions the female is experiencing rather than on its genetic quality because traits like the grey colouration of the rump show weak genetic variation (Kim et al., 2013). On the other hand, the width of the black tail band was found to have significant heritability and to be positively correlated with the individual body mass (Kim et al., 2013). López-Idiáquez et al. (2016b) found that the size, but not the number, of black spots was positively correlated with the body mass. On the other hand, females that bred earlier in the season had a lower number of spots, but they were larger compared to those of late-breeding females. Both the number and size of spots were not related to the clutch size, the number of fledglings generated or to the individual age. In another experiment, López-Idiáquez et al. (2016a) found that the rump melanic colouration might be a badge of social status in pre-laying female kestrels because kestrels attacked decoys with brown rumps significantly more than decoys with grey rumps. Overall, this work showed that melanic traits might reflect different qualities of the female that are important in dominance hierarchy. However, it is unclear to what extent melanic traits of females are shaped by both sexual and social selection.

Carotenoid-based colourations also appear to convey information about some qualities of the female. For example, female skin colouration and circulating non-enzymatic antioxidant capacity both increased across the mating to the nestling-rearing period (Casagrande et al., 2011). It is, however, difficult to explain this temporal covariation in the context of the sexual selection theory because the signalling function of carotenoid-based ornaments might be less important in kestrels during the nestling-rearing period than during the mating period. In kestrels, as in other birds of prey, males are responsible for provisioning the female with food during the incubation and both the female and offspring for around two weeks after hatching. Thus, it might be that males adjust their hunting effort to the quality of the female, which is conveyed by its skin colouration. If brighter females have more non-enzymatic antioxidants, it might be that they can deposit more antioxidants into the eggs, which are very important for the development and survival of the embryo (Surai et al., 1999).

5.8.2   Egg Pigmentation as an Extended Female Phenotype

Eggshell pigmentation varies within and among clutches (Figure 5.7). One reason for such variation might lie with the potential role of eggshell pigmentation as a sexual signal external to the body (Moreno & Osorno, 2003; *extended phenotype hypothesis*, Dawkins, 1982). Pigments deposited by mothers into the eggshell have several potential functions, such as protection against pathogens or pro- and antioxidant activity. They may also provide structural overcompensation for shell thinning due to a deficit of dietary calcium (Williams et al., 1994; Afonso et al., 1999; Shan et al., 2000; Gosler et al., 2011). Eggshell colouration might therefore signal the female's state to the male, which in turn would adjust its reproductive investment. In other words, according to this hypothesis, males would provide more parental care where eggs deposited by their mates are particularly rich in pigments, thus indicating that both the female and her eggs are of high quality.

Chromatographic analyses identified at least two pigments in the eggshell of kestrels: protoporphyrin IX at a concentration of 58 ($\pm$ 45 standard deviation) nmol g^{-1} of eggshell and biliverdin IXα at a concentration of 0.02 ($\pm$ 0.01 standard deviation) nmol g^{-1} of eggshell (Fargallo et al., 2014). However, effective engagement of eggshell colouration in sexual signalling has not been demonstrated in kestrels (i.e. the capacity of males to discriminate the female's state from egg colourations). Martínez-Padilla et al. (2010) suggested that the egg colouration is not a postmating sexual signal in species such as the kestrel, where males determine the female's state at the time of egg laying by providing food to them. Moreover, preliminary experimental investigations found that a reduction in pigment concentration in the eggshell did not have any detrimental consequences for the egg hatchability or nestling survival (Fargallo et al., 2014), suggesting that the pigmentation has low fitness value.

Figure 5.7 Eggs vary significantly in pigmentation in the common kestrel, but the meaning of this variation has not yet been understood.

5.9   Copulatory Behaviour

Male kestrels invest heavily in parental care; thus, selection is expected to favour the evolution of behavioural strategies that enable males to reduce any risks of extra-pair

Figure 5.8 Copulation between mates during the pre-laying period. Photograph by David Costantini.

copulations of their mates. Behavioural strategies can include frequent copulations in the pre-laying period (Figure 5.8), mate-guarding and aggressive behaviour against other males. If extra-pair copulations occur at a low frequency, behaviours that increase paternity probability would also be displayed at low frequencies. However, the male quality is also important because females mated with high-quality males would rely on extra-pair copulations less than those mated with low-quality males.

Vergara et al. (2007) found that high-quality males (e.g. older and more experienced individuals) were less aggressive against male intruders than low-quality males, indicating that they might have perceived a lower risk of extra-pair copulations of their mates than that perceived by low-quality males. This hypothesis was further supported by a work showing that high-quality males spent less time in copulatory behaviour than low-quality males (Vergara & Fargallo, 2008a). Also, the frequency of copulations was not linked to any traits of male quality, indicating that high-quality males did not compensate for a shorter copulation duration by performing more copulations.

The *paternity assurance hypothesis* does not seem to fully explain the copulatory behaviour of kestrels. Considering that kestrels show an early peak of copulations outside the fertile period of females (common kestrel in Meijer & Schwabl, 1989; lesser kestrel in Negro et al., 1992; American kestrel in Villarroel et al., 1998) and copulation attempts may occur even in winter (Village, 1990), it can be concluded that the function of copulatory behaviour might not be simply fertilisation and paternity assurance. For example, it has been suggested that frequent copulations in the lesser kestrel might be related to pair bonding and sexual stimulation of pair members (Negro et al., 1992). In contrast to colonial species like the lesser kestrel, in territorial and solitary breeder species like the common kestrel, copulations might also be used to signal territory ownership (Negro & Grande, 2001). However, Zink (1998) found that the frequency of within-pair copulations increased with breeding density, particularly during the peak of fertility. Thus, the meaning of copulatory behaviour might vary across conditions and species (e.g. Zink, 1998; Negro & Grande, 2001; Ille et al., 2002). What is certainly true is that extra-pair paternity is low in kestrels compared to other bird species. A study on common kestrels in Finland found that the extra-pair paternity varied from 0% to 5.4% across three reproductive seasons (Korpimäki et al., 1996). In lesser kestrels, the extra-pair paternity varied from 3.4% to 8.3% among various populations (Negro et al., 1996; Alcaide et al., 2005). In American kestrels, the extra-pair paternity was slightly higher, amounting to 11.2% of 89 nestlings examined (Villarroel et al., 1998).

5.10 Conclusions

We have learned much about the proximate causes of individual variation in body colourations of kestrels. The biochemical bases of both plumage and skin colourations have been identified. However, we still lack a clear understanding of the endogenous mechanisms that regulate the expression of colourations and what information about the individual state is conveyed by the different metrics of body colourations.

What is also critically lacking so far is a clear demonstration of which male traits are important in mate choice. There is a need for robust experimental settings that can test the complementary or antagonistic roles of body colourations and individual morphology (e.g. wing length) in mate choice. It is very important to consider carefully that plumage and skin colourations differ significantly in temporal dynamics and the phase of the life cycle during which they are produced. Thus, the lack of correlation between these traits might mean (i) that they convey different information about an individual's state or (ii) simply that intrinsic differences among traits in dynamics make them hardly comparable without adequate standardisations.

The ultraviolet component of colourations is also rarely considered, but it seems to be important in influencing the mate choice (Zampiga et al., 2008b). We need experiments with larger sample sizes that are replicated across different environmental conditions and that test simultaneously the interactions among multiple secondary sexual traits in driving the mate choice. Male kestrels also display specific behaviours during the mating and courtship periods that contribute to pair formation and bond consolidation. Thus, any experimental work that focuses on single traits would limit our capacity to infer about their real meaning in mate choice or in social competition. Recent theoretical research suggested that we should also consider the possibility that, under some circumstances, rather than choosing high-quality males (*choice of high-quality mates hypothesis*), females might use signals to avoid low-quality males (*avoidance of low-quality mates hypothesis*) (Gomes & Cardoso, 2018). This might help explain the high diversity of male sexual signals and the mechanisms driving the female's mate choice.

Body colourations were also associated with a number of behavioural and condition-dependent traits in young kestrels. It has also been proposed that colourations of females and of their eggs might have an important role in both sexual and social communication. Their meaning is, however, far from being clear, because a small number of studies have analysed the biological meaning of body colourations of both nestlings and females. This topic deserves further attention and will certainly be a fruitful area of investigation.

6   The Reproductive Cycle: From Egg Laying to Offspring Care

6.1   Chapter Summary

This chapter describes the breeding cycle of kestrels from the egg laying to the nestling-rearing phase. It illustrates the different reproductive strategies of males and females, the endogenous mechanisms and environmental conditions that affect the laying date and the clutch size, the adaptive meanings of egg volume and of hatching asynchrony, and the factors that affect the probability of nestlings surviving until fledging.

6.2   Introduction

Reproduction is a demanding phase of animals' lives, because they must produce, and in some cases protect and provision, their young. The reproductive activity of birds includes the phases of mating (Chapter 5), egg laying and parental care. When to start laying and how many offspring to generate are critical decisions. Females may optimise their reproductive strategies by adjusting the start of laying and the number of eggs laid to the prevailing environmental conditions or to their current state. The post-hatching period until fledging lasts around one month (Figure 6.1), during which partners swap to provide food to their offspring. The amount of parental care is affected by several factors, such as the current condition of parents, brood value, food availability and age of nestlings. Much work has been dedicated to characterise the causes and consequences of the phenology and success of reproduction in kestrels across time and contexts.

6.3   When to Lay

Understanding the causes and consequences of individual variation in the timing of reproduction is a major target of evolutionary ecology. The timing of reproduction is critically important for birds, particularly in those environments where conditions like food availability or weather conditions change with time. Natural selection has restricted the reproduction of birds in temperate zones to the period of the year when food is more abundant, the daylength is longer and the weather conditions are more clement. Despite the strong selection for early breeding, the *individual optimisation model* predicts that each individual has an optimal breeding window. Thus, the higher

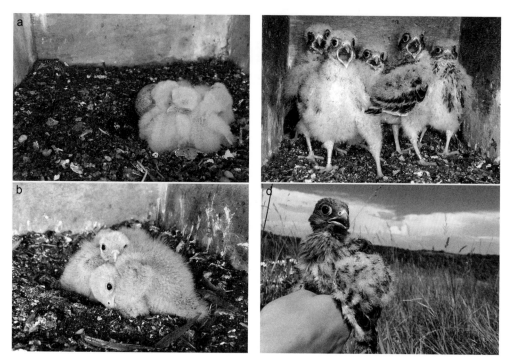

Figure 6.1 The post-hatching growth period until fledging lasts around 27–32 days in common kestrels. Images show nestlings of (a) one, (b) seven, (c) 15 and (d) 21 days of age. Photographs by Gianluca Damiani.

reproductive success that is usually recorded in early breeders compared to late breeders might be due to (i) better environmental conditions (e.g. food availability, weather conditions) at the beginning of the breeding season or (ii) higher-quality individuals breeding earlier than lower-quality individuals.

6.3.1   Hormonal Regulation of Laying Date

The egg-laying period of common kestrels in Europe is highly variable (Figure 6.2), ranging from mid-March to early June (Village, 1990; Costantini et al., 2010a, 2010b). Generally, egg laying occurs earlier in southern countries. Increasing daylength appears to be one important clue that switches the reproductive activity on. In temperate zones, daylength gets longer from autumn to spring, meaning that kestrels would be stimulated to breed during the season when food is abundant, but also weather conditions are better.

Experimental work carried out on captive American kestrels showed that individuals that had just bred were stimulated to breed again in midwinter by exposing them to artificially short daylengths earlier in autumn (Bird et al., 1980). Similarly, the stimulatory effect of daylength on the breeding behaviour of common kestrels occurs only after a period of short daylength, such as that experienced in the wintertime (Meijer, 1988, 1989; Meijer et al., 1990, 1992). Otherwise, kestrels would be

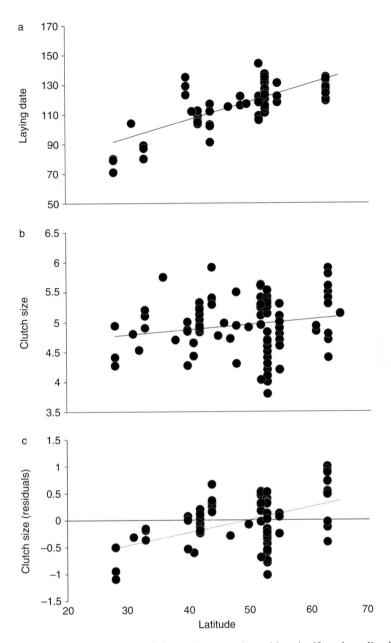

Figure 6.2 (a) Northern populations of common kestrel lay significantly earlier than southern populations ($p < 0.001$ in linear mixed models with nest type [natural vs. nest box] as a fixed factor and article as a random factor). (b) Clutch size is weakly associated with latitude ($p = 0.05$ in linear mixed models with article as a random factor). (c) The association between clutch size and latitude was highly significant when taking laying date into account ($p < 0.001$ in linear mixed models with nest type and laying date as fixed factors and article as a random factor). The panel shows residuals of clutch size that were obtained by a linear regression of clutch size onto laying date. Laying date is expressed as days elapsed from 1 January. (Data were collected from: Heim de Balsac & Mayaud, 1962; Cavé, 1967; Rockenbauch, 1968; Hagen, 1969; Géroudet, 1978; Nore, 1979; Kuusela, 1983; Pikula et al., 1984; Bonin & Strenna, 1986; Bergier, 1987; Riddle, 1987; Meijer et al., 1988; Hasenclever et al., 1989; Kostrzewa, 1989; Village, 1990; Aparicio, 1994b; Plesník & Dusík, 1994; Gil-Delgado et al., 1995; Daan et al., 1996; Korpimäki & Wiehn, 1998; Fargallo et al., 2001, 2002; Salvati, 2002; Carrillo & González-Dávila, 2005, 2009, 2010a, 2010b; Martínez & Calvo, 2006; Charter et al., 2007a, 2007b; Costantini et al., 2010a, 2014; Kaf et al., 2015; Kreiderits et al., 2016; López-Idiáquez et al., 2016a.)

insensitive to daylength (*photorefractoriness*). This scenario is likely different in tropical regions, where daylength is fairly constant throughout the year. The question then is which endogenous and environmental factors regulate the transition from a non-reproductive to a reproductive status in tropical regions.

Meijer and Schwabl (1989) examined hormonal changes of captive and free-ranging female common kestrels to investigate if birds breeding early, late, or not at all differed with respect to levels of the luteinising hormone (LH; a hormone produced by gonadotropic cells that triggers ovulation) and corticosterone, and if food intake affected the hormonal cycle. LH levels rose during courtship, reached a maximum during egg laying, dropped to half of the maximum during incubation and declined to winter levels during the nestling-rearing phase (Meijer & Schwabl, 1989). In captive females, LH levels were already elevated during winter compared with free-living birds. In captive females, corticosterone levels increased during courtship and reached a maximum during egg laying as previously observed in American kestrels (Rehder et al., 1986). After incubation had started, levels of corticosterone decreased. Changes in plasma levels of LH and corticosterone were much smaller and maximum levels were significantly lower in non-laying than in laying females. During the period before courtship occurred, LH levels were similar among early-laying, late-laying and non-laying females (Meijer & Schwabl, 1989). From the courtship phase onward, LH increased slightly in both late-laying and non-laying females, while it increased sharply in early-laying females.

The next question is which environmental factors may stimulate LH production. Captive kestrels undergoing a food restriction experiment did not show any courtship behaviour and had a significant decrease in body mass (Meijer & Schwabl, 1989). Courtship behaviour reappeared and body mass increased just 2–3 days after food was provided *ad libitum* to the kestrels. Interestingly, LH levels in food-rationed females increased similarly to those in females fed *ad libitum*, while corticosterone levels were much lower in food-rationed females than in females fed *ad libitum*. Food provisioning caused an increase of corticosterone levels in food-rationed breeding and non-breeding females compared to levels recorded in *ad libitum*-fed females (Meijer & Schwabl, 1989).

Overall, these results suggested that at the start of the breeding season, female kestrels have functional reproductive systems that are sensitive to changes in daylength and are only slightly affected by the availability of food. Thus, the among-female variation in the timing of laying (early vs late) would not be explained by a modulation mediated by corticosterone or LH; rather, they might be stronger linked to seasonal changes in the availability of food (Meijer & Schwabl, 1989) or to other factors. The question then is which hormones link food availability to the female's reproductive phenology.

6.3.2   Role of Food Availability

Earlier work by Dijkstra et al. (1982) showed that food availability might have a significant effect on the reproductive decisions of females. They supplied free-living kestrels with dead laboratory white mice prior to egg laying. Daily rations given to a pair varied between 100 and 120 g, corresponding to about twice the daily metabolism of kestrels in captivity (Kirkwood, 1979). Those kestrels that were given extra food laid earlier than those that were not in years with scarcity of prey in the environment, while the effect of food supplementation on the laying date seemed to be negligible when prey were abundant (Dijkstra et al., 1982), suggesting a threshold below which a limited supply of food has a significant impact on the female's condition and hence on her propensity to reproduce. These preliminary results were corroborated by a more detailed study on the same kestrel population in the Netherlands (Meijer et al., 1988). In the first half of the breeding season, kestrels that received extra food advanced significantly the laying date during low but not high vole density years. Conversely, in the second half of the breeding season, the effect of food supplementation on the laying date was significant, but weaker, and it was similar between years with high and low vole densities (Meijer et al., 1988). Moreover, captive female kestrels did not lay as long as experimental food rationing lasted (Meijer et al., 1988). Similarly, in work carried out in Spain, food supplementation stimulated kestrels to lay earlier, particularly in those females that bred early in the season (Aparicio, 1994b). A link between food availability and the timing of reproduction was also observed in the lesser kestrel. Aparicio and Bonal (2002) observed that food-supplemented lesser kestrels bred earlier than unsupplemented lesser kestrels irrespective of the breeding habitat.

6.3.3   Endogenous Constraints

Food supplementation experiments suggested that female kestrels might respond to food availability by shifting their laying date forward or backward in the season. This would indicate a certain degree of plasticity of the laying date. Data collected from free-living kestrels showed that the within-individual repeatability of laying date over seasons was low in a Dutch population, suggesting that kestrels may significantly advance or delay egg laying without much endogenous constraint (Meijer et al., 1988). Further work suggested that females might change the laying period within a given temporal window, which appears to be endogenously determined and affected by the daylength (Meijer, 1988; Meijer et al., 1990).

Although these results supported a significant role of food availability in determining the laying date of kestrels in the Netherlands, work carried out on other kestrel populations found that the laying date was highly repeatable, suggesting low plasticity for this trait (Finland: Korpimäki & Wiehn, 1998). Moreover, a number of studies carried out on other bird species suggested that the laying date might be heritable (van Noordwijk et al., 1981; Hendry & Day, 2005; Catry et al., 2017). Thus, the laying date might not only depend on the functional responses of the reproductive system to the environment, but also on some genetic programming that would lead to reduced gene flow between early- and late-breeding individuals.

A study on the population genetics of common kestrels breeding in central Italy found a reduced gene flow between early and late sympatric breeders (Casagrande et al., 2006a). A comparison of five microsatellite loci between early- and late-hatched nestlings (in other words between their early- and late-breeding parents) showed that

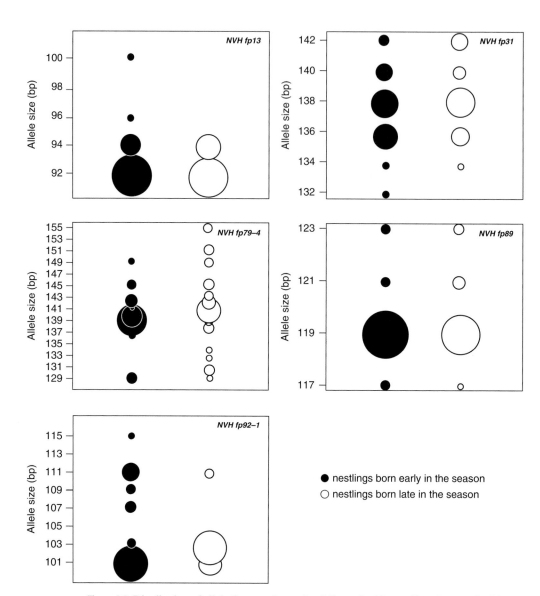

Figure 6.3 Distribution of allele frequencies at the different loci in nestlings born early (black) and late (white) in the breeding season. Distributions were significantly different across all five microsatellite loci between early- and late-born nestlings. Reprinted from Casagrande et al. (2006a).

(i) the number of alleles per locus ranged from 3 to 12, but the mean number of alleles per locus was similar between them, (ii) several alleles were unique to either early- or late-hatched nestlings (e.g. loci NVH fp79-4 and NVH fp92-1), (iii) significant departures from the Hardy–Weinberg equilibrium were not observed between them, but were evident within each group of nestlings, (iv) the genetic variability was similar between them and (v) there was a highly significant genetic differentiation between the two groups (Figure 6.3).

The nestlings included in this study came from broods born at the beginning or end of the breeding period, and were separated in time by at least 30 days. The genetic differences were, therefore, most likely due to differences in breeding time. As the microsatellite loci are neutral with respect to natural selection, the genetic differences between early and late breeders indicated reduced gene flow due to (i) genetic drift between partially reproductively isolated breeding groups or (ii) linkage between microsatellite loci and loci under differential selection between early and late breeders. These results also suggested that this genetic differentiation between early and late breeders was due to assortative mating-by-time (Casagrande et al., 2006a). Thus, individuals breeding at the extremes of the season produce offspring that would breed at similar times. In other words, the progeny of early breeders would mate with greater probability with the progeny of other early breeders, whereas the progeny of late breeders would mate with the progeny of other late breeders. This would allow genetic differences to increase with respect to breeding time. The possibility of assortative mating-by-time for kestrels is further strengthened by the very low incidence (below 5%) of extra-pair paternities (Section 5.9) that were recorded in several kestrel populations (Finland: Korpimäki et al., 1996; Spain: Vergara et al., 2015). Early studies on kestrels carried out in the Netherlands suggested that early breeders have adaptive advantages compared to late breeders because they can produce larger and healthier broods, thereby achieving greater reproductive success (Daan & Dijkstra, 1988; Daan et al., 1990). Moreover, it has been suggested that male kestrels hatched early in the season may recruit in the breeding population earlier than male kestrels hatched late in the season (Daan et al., 1990). However, these studies have disregarded possible adaptive advantages of breeding late as an alternative strategy to breeding early (Price et al., 1988). It might be that breeding early or late are two different strategies evolved to maximise reproductive outcomes of the population (Section 6.4). This might be particularly plausible for kestrels breeding at lower latitudes due to a long breeding season compared to kestrels breeding at higher latitudes. However, these results do not prove a genetic regulation of laying date.

6.3.4 Laying Date and Weather Conditions: An Example of Flexibility

If we assume that the laying period is endogenously constrained, it still has to harbour a certain degree of flexibility to enable the female to adjust it to the current environmental conditions in order to start reproducing when they are closer to optimality. Costantini et al. (2010a) found that females delayed egg laying in rainier and colder breeding seasons across a study period of 10 years in central Italy (Figure 6.4). At the population level, there was a delay of approximately one day for an increase of 10 mm of rainfall during early spring.

A study on the impact of weather conditions on the laying date recorded in 23 different areas within 28–65°N and 17–28°E also found that there was a delay of egg laying during rainier and colder breeding seasons (Carrillo & González-Dávila, 2010a). Although the effect of temperature on the timing of reproduction may have been indirect, for example via food phenology, research on passerine

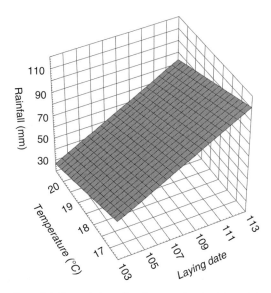

Figure 6.4 Kestrels breeding in central Italy delayed their laying date in rainier ($r = 0.82$, $p = 0.003$) and warmer ($r = -0.71$, $p = 0.02$) springs (March–April). Annual mean values of laying date are shown. Laying date is expressed as days elapsed since 1 January.

birds showed that temperature might itself have a direct effect on the timing of breeding (Visser et al., 2009).

The effect of weather conditions on the laying date might vary across environments, such as temperate or semi-arid regions. A study carried out in the semi-arid ecosystems of Tenerife found that heavy rainfall in the previous autumn was associated with an earlier start of the egg-laying period and a longer breeding season (Carrillo & González-Dávila, 2010b). In contrast, rainfall during the reproductive season did not affect the laying date. Carrillo and González-Dávila (2010b) suggested that winter rainfall might have favoured plant growth and flowering, increasing local abundance of prey, such as lizards.

6.3.5 Unexplored Endogenous Factors

Other endogenous factors might affect the female's reproductive decisions. For example, an experiment carried out on captive canaries (*Serinus canaria*) showed that females, whose molecular oxidative damage was experimentally increased by the suppression of the antioxidant glutathione before mating, significantly delayed the start of egg laying (Costantini et al., 2016). These results suggested that birds might also need to balance their reproductive decisions against cellular oxidative stress in order to optimise their lifetime reproductive success. This may occur through the suppressive effects that oxidative stress may have on the reproductive hormones that regulate the transition from a non-reproductive to a reproductive status (Agarwal & Allamaneni, 2004). Adjusting the start of egg laying to the current oxidative stress level would have several advantages for the female because oxidative stress can have detrimental effects on fertility or embryo development and survival (reviewed in Costantini et al., 2016).

6.4   Incubation Behaviour and Clutch Size

Conversely to laying date, the decision of how many eggs to lay is under stronger female control. Kestrels usually lay four to six eggs once a year. Rarely, they lay a second clutch if the first fails or is not completed (Village, 1990). The clutch size generally decreases with laying date and larger clutches are laid in northern regions independently from the laying date (Figure 6.1).

6.4.1   Follicle Development and Incubation Behaviour

Follicles from which eggs develop in the ovary are formed during the winter, but they develop in around nine days with an increase in the growth rate in the last week before laying the egg (Meijer et al., 1989, 1990, 1992). The regulation of clutch size is accomplished through the resorption of developing follicles, which appears to be affected by the number of eggs already laid (Meijer, 1988; Meijer et al., 1990). Stimuli from the eggs (tactile, visual) are important in enhancing incubation behaviour and, thus, in stimulating clutch fixation by follicle suppression (Beukeboom et al., 1988; Meijer, 1988; Meijer et al., 1990).

Kestrels lay eggs at intervals varying between one and four days with an interval of around two days being the most common (Village, 1990; Aparicio, 1994a; Wiebe et al., 1998a). Experiments carried out on kestrels in the Netherlands showed that, in the course of egg laying, the responsiveness of females to experimental removal of the egg is switched off at some point, which occurs on average four days before the female lays her last egg (Beukeboom et al., 1988). At this point, the female seems to be no longer able of developing extra follicles to produce new eggs in order to replace those that were artificially removed. As previously shown in the American kestrel (Porter, 1975), the decision to terminate laying in common kestrels is taken sooner in late breeders than in early breeders, probably because (i) late breeders develop incubation behaviour more rapidly than early breeders, (ii) late breeders start laying at a higher incubation tendency, which would result in faster follicle suppression and laying of smaller clutches or (iii) late breeders are on a stricter breeding schedule (Beukeboom et al., 1988; Meijer, 1988; Meijer et al., 1990). The link between incubation behaviour and termination of egg laying is also affected by other factors (Section 6.6).

6.4.2   Hormonal Regulation

The decline of clutch size with laying date might be due to a slow progressive switching off of the female reproductive system. A question then is which endogenous mechanisms would synchronise the maturation of the reproductive system with the daylength. The hormone prolactin is one major candidate. In birds, prolactin rises gradually in spring before laying and quickly during laying (Dawson & Goldsmith, 1982, 1985; Hall & Goldsmith, 1983). Prolactin secretion is enhanced by tactile stimulation from the eggs through the brood patch (Hall & Goldsmith, 1983), i.e. the patch of featherless skin on the underside of birds that makes it possible to transfer heat to their eggs while incubating. Experimental administration of prolactin to females suppressed the production of gonadotropins, which are the hormones that stimulate the production of eggs (El Halawani et al., 1986, 1991).

Work carried out on captive birds that went through the normal reproductive cycle also supported the role of prolactin as one regulator of incubation behaviour in common kestrels (Meijer et al., 1990). This work showed that plasma prolactin (i) increased significantly from February to early June, (ii) was lower in early- than late-breeding females before egg laying and (iii) was similar between early- and late-breeding females while laying (Meijer et al., 1990). A significant role of prolactin in the regulation of incubation behaviour was later demonstrated experimentally in the American kestrel (Sockman et al., 2000). However, this study did not find support for the role of prolactin in regulating the clutch size. Although a negative relationship between plasma prolactin during clutch formation and clutch size has been found, the administration of prolactin to females did not affect the clutch size (Sockman et al., 2000).

6.4.3   Optimal Clutch Size–Laying Date Combination Hypothesis

In kestrels, as well as in other single-brooded birds, the clutch size declines with the laying date. Two main hypotheses have been proposed to explain the factors responsible for such a decline. The *optimal clutch size–laying date combination hypothesis* postulates that the clutch size is determined by factors (e.g. daylength) that are independent from the food availability (Drent & Daan, 1980; Daan et al., 1989a, 1990). Birds would adjust their clutch sizes to the optimum size for the laying period and any changes in food supply would not affect the relationship between laying date and clutch size. This hypothesis emphasises the role of two main variables: the reproductive value of the egg, which depends on the laying date, and the number of feedable offspring, which depends on territory or parental quality. The reproductive value of the egg is predicted to decline with the season, while the number of feedable offspring should increase and then decrease with the season. Thus, the optimal clutch size would result from a trade-off between egg reproductive value and food availability. Experiments of food manipulation in both free-living and captive kestrels in the Netherlands supported the model of Daan et al., indicating that the availability of food has no impact on the clutch size (Meijer et al., 1988). Thus, food abundance would affect clutch size only through its effect on the date of laying (Daan et al., 1990; Meijer et al., 1990). Age also did not affect the clutch size (Meijer et al., 1988). In lesser kestrels, Catry et al. (2017) also found that the seasonal decline in clutch size was due to conditions experienced before the start of the laying period rather than to the availability of food later in the season.

There are several caveats in the optimal clutch size–laying date combination hypothesis. For example, in seasonal environments, the peak of prey availability is not constant, but varies among years. Thus, if the clutch size is really dependent on the laying date, then the female might lay a number of eggs that is not optimal for that given period if food availability is low in that particular year. Similarly, short unpredictable events (e.g. bad weather, delayed arrival at the breeding grounds) might delay the beginning of egg laying without affecting the availability of prey during the nestling-rearing period.

6.4.4   The Food Limitation Hypothesis

The *food limitation hypothesis* postulates that it is the availability of food that affects the clutch size independently from the laying date (Newton & Marquiss, 1981; Aparicio, 1994a, 1994b). Thus, if a female lays in the same period for two consecutive years, she will lay a larger clutch size in the year when there is more food available. This would mean that the number of potentially feedable offspring is adjusted to the current conditions of food abundance. Conversely to the results obtained by Meijer et al. (1988), studies carried out on other kestrel populations found support for the food limitation hypothesis. In south-west Scotland, a strong correlation has been found between clutch size and abundance of voles (Shaw & Riddle, 2003). In Spain, the clutch size was significantly larger in kestrels that received extra food than in those that did not, even if there were no differences in laying date between them (Aparicio, 1994b). Moreover, the clutch size declined with laying date only in those kestrels that were not supplemented with food (Aparicio, 1994b). The effect of food supply on clutch size was at least partly explained by the laying interval, with fed females (thus in better nutritional status) that reduced the time elapsed from the laying of one egg to that of the next one. Similarly, work carried out on common kestrels in Finland found that females given supplementary feeding before and during the egg laying produced larger clutches than unfed control females without affecting the laying period (Korpimäki & Wiehn, 1998). Food supplementation increased the number of offspring produced in years of both low and high availability of prey, indicating that food may limit the reproductive decisions of females even at high natural levels of prey (Lack, 1947, 1954). Thus, females might shift the laying date to some degree in order to adjust the reproductive activity to the availability of prey. This is particularly important in those environments where densities of the main prey show marked differences among years. Implicit in the food limitation hypothesis is that the reproductive value of eggs for the same laying date also varies among years depending on the local abundance of prey.

6.4.5   The Individual Optimisation Hypothesis

Another question is whether the optimal time for breeding differs or not among pairs; in other words, if there is an individual optimisation of breeding time. To test this hypothesis, Aparicio (1998) manipulated the hatching date by cross-fostering clutches of equal size but differing in laying dates. This experiment showed that experimentally advanced clutches produced fewer fledglings and fledglings of poorer body condition than control clutches, thus providing some support for the *individual optimisation hypothesis* (Aparicio, 1998). In a Finnish population of kestrels, it was found that although the number of eggs laid may be adjusted to the food available while laying, it might not match the food available later in the season (Wiebe et al., 1998a). The choice of laying many eggs might turn maladaptive if, for any reason, prey availability collapses unpredictably with time. It is, therefore, important to give careful consideration to both current food abundance and predictability of food availability in the future.

Weather conditions may be important in determining both reproductive decisions and success. A 10-year study carried out in the Mediterranean region found that kestrels laid smaller clutches after warmer and drier winters (Costantini et al., 2010a). Conversely, weather conditions during the early phases of the reproductive season did not affect mean clutch size or the variation in clutch size among pairs. Both the mean clutch size and its skewness (with laying period controlled for in statistical analyses) were negatively correlated with the winter temperature and the North Atlantic Oscillation (NAO), an index of non-local climatic conditions (higher values of NAO index indicate drier conditions in the Mediterranean region). Conversely, the standard deviation of clutch size correlated positively with the NAO index, meaning that dry winter conditions increased variation of clutch sizes in the population (Costantini et al., 2010a). These results seem to be against predictions because it could be expected that kestrels lay larger clutches after mild than harsh winters. One possible explanation is that mild winters shifted the phenology of prey populations and kestrels were unable to adjust their breeding time to the changes in food availability as suggested by the lack of association between winter weather and laying date (Section 6.3). As a consequence, kestrels could have reduced their investment in reproduction. Another explanation is that favourable winter conditions (e.g. higher availability of food; Ashmole, 1961) increased survival and reproductive perspectives of low-quality individuals or of young that produce smaller clutches, consequently reducing the mean clutch size and increasing the variation in clutch size of the population, respectively (Costantini et al., 2010a). A similar hypothesis was proposed to explain the great number of small clutches observed after favourable winters in pied flycatchers (*Ficedula hypoleuca*), because mild winters may increase food availability, so low-quality individuals have higher probabilities of survival (Laaksonen et al., 2006). This hypothesis is further supported by a comparative study of clutch sizes across 23 different geographic areas that found a reduced mean clutch size after winters rich in resources (Carrillo & González-Dávila, 2010a). However, the effects of weather conditions on clutch size are not always evident. For example, a study on common kestrels breeding in the urban areas of Vienna did not find any evidence for an effect of weather on clutch size (Kreiderits et al., 2016).

6.5   Egg Volume

Egg volume is one of the most widely studied life-history traits of birds (Williams, 1994; Christians, 2002). Although many sources of within- and between-species variation in egg volume (e.g. food availability, female body condition, female age, genetics) and its consequences for offspring (e.g. body size at hatching, growth pattern, survival) have been identified, the selective forces that maintain such variation remain not fully understood yet (Williams, 1994; Christians, 2002).

In kestrels, egg volume is very variable (see Figure 5.8). Valkama et al. (2002) examined the variation among clutches in the egg volume of common kestrels in western Finland over a period of 12 years. They found evidence for a potentially significant contribution of genetic background in determining egg volume and a lack of clear benefits for fitness (no association with nestling mortality). The overall mean egg volume was 19.5 cm^3 and varied from 16.1 to 25.3 cm^3. Egg volume was highly and significantly repeatable ($r = 0.88$; repeatability describes the upper limit of heritability according to Falconer, 1981) in females over time. Moreover, the mean egg volume of the second breeding attempt of individual females was strongly correlated ($r = 0.87$) with the mean egg volume of their first breeding attempt within the same breeding season, independently from any changes in food supply, female body condition and laying period (Valkama et al., 2002). The egg volume decreased with laying date in the years with low abundance of voles or in the years with a decrease in the abundance of voles with laying date. Conversely, in those years with high abundance of voles, egg volume was not significantly associated with the laying date (Valkama et al., 2002). Similar results were obtained in the American kestrel (Wiebe & Bortolotti, 1995). However, the decline in egg volume with laying date was not affected by food supplementation in captive American kestrels, suggesting that seasonal changes in egg volume might not be strongly associated with the supply of energy or nutrients (Bird & Laguë, 1982b).

In a Mediterranean population of common kestrels, the covariation between egg volume (mean $\pm$ standard deviation, range: 19.5 ± 1.6 cm^3, 12.5–25.4 cm^3, $n = 1368$) and laying period varied across years (Costantini et al., 2010b). In one year, there was a low positive correlation between egg volume and hatching date, whereas the correlation between egg volume and hatching date was negative and quite variable in magnitude in the other study years. This negative correlation between egg volume and hatching date was mostly evident for the four-egg clutches, while it was not for both the five- and six-egg clutches (Figure 6.5; Costantini et al., 2010b).

Other studies found a clearer effect of food availability on the egg volume. Food supplementation prior to egg laying increased the egg volume of first, but not of middle and last, laid eggs within a same clutch in a population of common kestrels in Spain (Aparicio, 1999). In a Canadian population of the American kestrel, food supplementation had a stronger effect on the egg volume than that

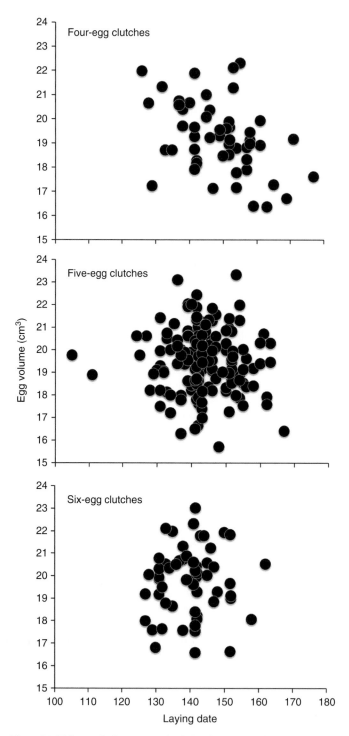

Figure 6.5 This graph shows a pool of clutches from different years. Although some variation between years occurred, the seasonal decline of egg volume was more evident in four-egg clutches ($r = -0.34$, $p = 0.003$) than in five- or six-egg clutches (Costantini et al., 2010b).

found in common kestrels (Wiebe & Bortolotti, 1995). Such an effect was caused by supplemented females laying wider, but not longer, eggs than those laid by control females (Wiebe & Bortolotti, 1995). In this context, it is important to consider both the energy reserves accumulated before egg laying starts and the daily food intake while laying. Accumulated energy reserves might be quickly depleted for the production of the first eggs; thus, the egg volume of the last laid eggs might be more dependent on the current availability of food for the female. As a consequence, the effect of an experimental supplementation of food might vary if it is done before the start of egg laying or during egg laying. In an experiment on the American kestrel, food supplementation carried out until the deposition of the first egg affected the egg volume similarly to food supplementation carried out throughout the whole laying (Wiebe & Bortolotti, 1995). These results suggest that energy reserves accumulated before laying (capital) might be more important in determining the among clutch variation in egg volume than the energy intake during laying (income; Wiebe & Bortolotti, 1995; see also Drent & Daan, 1980). Conversely to the variation among clutches in egg volume, the within clutch variation in egg volume was weakly affected by food supplementation (Wiebe & Bortolotti, 1996). Moreover, it was not associated with the local abundance of prey or with the body condition of both males and females. If the within-clutch variation in egg size has an adaptive meaning, it would be expected that such variation is higher in large clutches when food shortages during the nestling period would have been more likely to occur; this would cause a body size hierarchy that increases chances of survival of the biggest nestlings. The experimental results presented by Wiebe and Bortolotti (1996) did not support the hypothesis that female American kestrels can vary the egg size across the laying sequence depending on the current availability of food, thus rejecting the hypothesis that the egg volume has some adaptive value. In line with this conclusion, Costantini et al. (2010b) found that hatched (mean ± standard error: 19.0 cm^3 ± 0.1) and unhatched (19.1 cm^3 ± 0.2) eggs belonging to a same clutch did not differ significantly in volume.

Another aspect of egg volume deserves further attention. The egg volume is the result of its length, width and shape. These parameters may provide valuable information to further our understanding of the way a female can modify the egg volume. In a population of kestrels breeding in central Italy, the length and width of an egg were generally correlated ($r = 0.55, p < 0.001$); however, the variation in egg length (mean ± standard deviation, range: 39.1 ± 1.5 mm, 33.5–45.2 mm; $n = 1368$) was higher than that of egg width (31.2 ± 1.0 mm, 27.0–34.3 mm; $n = 1368$; Costantini et al., 2010b). A plausible reason is that the egg width is constrained by the size of the female oviduct (e.g. Grant, 1982; Jàrvinen & Vaisanen, 1983). Thus, the egg length might be the most flexible trait that a female common kestrel can modify to regulate the egg volume (e.g. Grant, 1982; Jàrvinen & Vaisanen, 1983). This does not seem to be the case for other kestrel species. In a study on the American kestrel, the variation among clutches in egg volume was mainly due to differences in egg width rather than in egg length (Wiebe & Bortolotti, 1995).

6.6   Hatching Asynchrony and Hatching Success

The decision of when to start incubating appears to be under strong selective pressure because it may determine the success of the reproductive activity. Females can actually control the hatching asynchrony (difference in hatching time between the first and last hatched nestling) by using different incubation schedules. Data collected with thermometers placed in nests of common kestrels to record the female's body temperature (i.e. the presence of the female in the nest) showed that the onset of incubation varied significantly among females and increased gradually during the egg-laying phase (Beukeboom et al., 1988; Wiebe et al., 1998b).

Lack (1954) proposed that the hatching asynchrony would generate a body size hierarchy with adaptive value when abundance of food is low (*brood reduction hypothesis*) because it could increase the reproductive success. The larger first-born nestlings would outcompete the smaller last-born siblings and, as a result, would grow faster and have lower probability of dying. On the other hand, synchronous hatching when food occurs in limited supply would mean higher sibling competition and lower survival chances. Thus, the hatching asynchrony is expected to have a number of effects on offspring development and reproductive success (Section 6.6).

In Finland, the hatching span of kestrels ranged from 0 to 10 days, with most eggs within a same clutch hatching within 2–3 days (Wiebe et al., 1998a). While controlling for clutch size, hatching was more synchronous when there was some food shortage (i.e. during the low phase of the vole cycle) compared to when food was more abundant (Wiebe et al., 1998a). This result is seemingly in contrast with the predictions of the brood reduction hypothesis. However, the low phase of the vole cycle was characterised by more stable densities of prey compared to the high phase. This result suggests that kestrels might have regulated the hatching asynchrony in relation to the predictability of vole abundance and not to the total abundance of prey (Wiebe et al., 1998a). In other words, a low fluctuation in the abundance of voles would give the kestrels more reliable information about the availability of prey than the total abundance of voles during the breeding season. This strategy might be advantageous because numbers of voles may increase or even crash at different rates during the breeding period of kestrels, making the total amount of voles a poor indicator of their availability (Wiebe et al., 1998a).

Wiebe et al. (1998b) showed that female kestrels might also use different incubation schedules in order to regulate hatching patterns. Females may vary the onset of incubation during the laying phase and/or the rate at which the time allocated to incubation is increased during laying (Wiebe et al., 1998b). The percentage of time allocated to incubation can increase rapidly with the number of eggs laid until a threshold is reached (*rising incubation*; Wiebe et al., 1998b) or can change following a steady or pulsed pattern (*partial incubation*; Wiebe et al., 1998b). The rising incubation would result in almost synchronous hatching of most eggs from the same clutch. Conversely, partial incubation would result in more constant (steady pattern) or unpredictable (pulsed pattern) hatching intervals.

This variation among incubation schedules suggests that (i) hormonal control might not be the only factor involved in regulating incubation behaviour and (ii) the need to incubate might not determine the termination of laying, as suggested by Meijer et al. (1990). Generally, the eggs hatch in the order they are laid, but some minor exceptions occur for the first-laid eggs (Wiebe et al., 1998b). Conversely to the common kestrel, the pulse pattern of incubation appears to be more common in the American kestrel, particularly for small-bodied females, and first-laid eggs do not always hatch first (Bortolotti & Wiebe, 1993; Wiebe & Bortolotti, 1993). It has been hypothesised that the hatching pattern is constrained by the body size in the American kestrel because the ratio of clutch volume to female body mass is higher than that of the common kestrel (about 11.6% vs 9.5%) or of other raptor species (Bortolotti & Wiebe, 1993; Wiebe & Bortolotti, 1993; Wiebe et al., 1998b). In common kestrels, Wiebe et al. (1998b) could not find any link between body size and hatching pattern, but they found that females with a pulse pattern of incubation were in poor condition after the end of laying. Given the large energy requirements of egg formation and incubation (Meijer et al., 1989), shortage of food while laying might affect significantly the incubation behaviour and, consequently, the hatching pattern (Wiebe et al., 1998b). Thus, females might use a number of environmental and endogenous clues to facultatively regulate the hatching asynchrony (Wiebe et al., 1998b). When and why a female would rely more on a given strategy or another is an open question.

Egg volume is another factor by which females may have some control on the hatching asynchrony and success. Aparicio (1999) found that the egg volume had a significant effect on the duration of the incubation, which was prolonged by 0.21 days per millilitre of egg volume. Female kestrels could synchronise hatching by producing last-laid eggs than are smaller than first-laid eggs and by postponing the onset of incubation until the egg laying is complete (Aparicio, 1999). However, a large last-laid egg might increase the survival probability of the nestling, while a small last-laid egg might make the nestling more vulnerable to starvation owing to competition with siblings (Clark & Wilson, 1981; Slagsvold et al., 1984). In a kestrel population in central Italy, hatched and unhatched eggs within the same clutch did not differ in volume (Costantini et al., 2010b). The mean egg volume of a clutch did not also predict the hatching success of the clutch itself or the fledging success of the brood. Hence, the egg volume was not a good predictor of reproductive success in this kestrel population. Conversely, bigger eggs had higher hatching probability in a Canadian population of American kestrel (Wiebe & Bortolotti, 1995) and in a Finnish population of common kestrel (Valkama et al., 2002). In the American kestrel, clutches with an infertile egg did not differ in egg volume from those in which all eggs hatched (Wiebe & Bortolotti, 1995). However, they were significantly bigger than the eggs from clutches with partial hatching failure, which in turn were bigger than those with complete hatching failure (Wiebe & Bortolotti, 1995). These results were obtained while controlling for variation in female body size. It is unclear, however, the extent to which female body condition may affect the hatching success through egg size because females in better condition tend to lay bigger eggs. For example, if a female in poor condition needs to take long breaks while incubating because the male is unable to provide enough food for her, small eggs might cool more quickly than big eggs even if the male incubates the eggs while the female is away. Small eggs might also have a lower amount of nutrients and other substances (e.g. maternal antibodies) because they are laid by females in poor condition (Wiebe & Bortolotti, 1995). Ambient temperature during the laying phase does not seem to be a concern for the hatching probability. Eggs may hatch even after six days left unattended and exposed to ambient temperature before the start of incubation (Wiebe et al., 1998b). Diikstra reported that eggs hatched after being kept in a refrigerator for one month since the day of laying (personal communication). However, these explanations might be context-dependent. At lower latitudes, environmental conditions are different from those that kestrels experience at higher latitudes. Thus, the adaptive value of a given strategy may vary across environments. Identifying the relative magnitudes of different selective forces and how they operate under different ecological circumstances remain open topics.

The sex of the embryo might be another factor important in determining the hatching pattern. Blanco et al. (2003a) showed that female embryos of common kestrel had a shorter embryonic period than male embryos, and this resulted in earlier hatching of females. It is unknown if this shorter embryonic period is due to a faster growth and development rate or to hatching at an earlier stage of development. Although hatching earlier may be advantageous to gain a high rank in the size hierarchy within the brood, fast growth may also carry metabolic and physiological costs that might have negative consequences in adulthood.

Mothers may modulate the embryo development or buffer any costs adopting different strategies of resource provisioning to eggs. This would also give the female some kind of pre-hatching parental control on post-hatching sibling competition. Blanco et al. (2003b) found some support for a potential capability of female kestrels to allocate resources differentially to eggs according to embryo sex and laying order. First-laid eggs bearing a female embryo were lighter than first-laid eggs bearing a male embryo, but no differences in mass were found for eggs laid in the middle or last orders. Moreover, in clutches with the first-laid egg bearing a male, the eggs laid subsequently in the laying sequence were smaller than the first-laid ones (Blanco et al., 2003b). The opposite pattern was found in clutches with the first-laid egg bearing a female. The adaptive meaning of this egg mass provisioning related to embryo sex and laying sequence was, however, unclear because it did not affect hatching asynchrony, hatching success, or fledging success (Blanco et al., 2003b).

Infertility of eggs may represent a main cause of hatching failure (e.g. Wiebe, 1996). The causes of infertility have not been investigated in detail so far. It might be due to low reproductive experience of the male or poor coordination between mates in breeding behaviour. The body condition of the female during incubation does not seem to be particularly relevant. In a study on Finnish kestrels, Jönsson et al. (1999) found that the body mass of incubating females that had all their eggs successfully hatched was similar to that of females with partial or complete hatching failure.

Another reason for egg infertility might lie with low sperm quality, which could be due to poor condition of the male. Spermatozoa are very sensitive to stress and, given that techniques are available to assess seminal characteristics in kestrels (Dogliero et al., 2016), analysing the causes and consequences of variation in sperm quality among males would be an important line of research to develop. In our study population in Italy, infertile eggs generally come from clutches of five or six eggs. Assessing the laying order of such eggs could help us to understand the causes of infertility. For example, first-laid eggs might be more at risk if there is a mismatch in mating coordination between mates; however, the first-egg laid could act as a clue to stimulate subsequent copulation with the female. Also, a female might deposit more contaminants previously accumulated in her body into the first-laid eggs (Chapter 8) and the contaminants might affect the embryo development and survival.

6.7   Parental Care of Nestlings

6.7.1   Prey Delivery Rates

Hunting and prey delivery to nestlings are time-consuming and demanding activities for kestrels. Prey size and availability and nestling age are important determinants of the time and energy allocated to hunting and nestling feeding. The male provides most of the food until the nestlings are in the middle of their growth period. During this period, the female has a central role in processing the prey and feeding the offspring.

Masman et al. (1989) found that one-week old nestlings that were hand-raised in the laboratory had an average food intake of 66.8 g day^{-1}, compared to 62.6 g day^{-1} for nestlings in the field in the Netherlands. In the same study area, the food intake was also higher in experimentally reduced broods (81 g day^{-1}) compared to that of control broods (61 g day^{-1}; Dijkstra et al., 1990b). In northeast China, Geng et al. (2009) estimated that the biomass of prey consumed by nestlings was 48.2 g day^{-1} in an average brood of 4.8 nestlings. In Spain, the mean prey biomass per hour over a period of four years for 12–14-day-old nestlings was estimated to be 26.4 ± 14.4 g, ranging from 3.2 to 83.9 g (Navarro-López et al., 2014). In a study in Norway, Steen et al. (2011) estimated that an average brood of 4.3 nestlings of 12 days of age (range: 8–28 days) would consume 18.3 g h^{-1}, while a nestling would consume on average 4.2 g h^{-1}. This estimate was equivalent to 67.8 g day^{-1}, given an average daily activity period of 16.1 h. The estimated delivery rate of prey items required to feed an average brood was 91 per hour if the kestrels had provided only insects, and 3.4, 1.9, 0.83 and 0.52 if they had provided only lizards, shrews, voles or birds, respectively (Steen et al., 2011). This corresponded to one prey delivery per 40 seconds if feeding solely on insects and one for about 18, 32, 75 and 120 minutes if feeding solely on lizards, shrews, voles or birds, respectively (Steen et al., 2011). Thus, prey size plays an important role in determining the number of foraging flights that parents have to engage in.

Mates differ in how they allocate prey to offspring. Sonerud et al. (2013) observed that males used to give larger prey to females, which in turn dismembered the prey and fed the nestlings. In contrast, males used to give smaller prey directly to the offspring, probably because nestlings can dismember and swallow small prey on their own. Sonerud et al. (2013) suggested that this different prey allocation strategy between sexes is dependent on a balance among the female's control of the allocation of food between her own need and that of the offspring and on the optimisation of the time the male allocates to hunting. It also should not be forgotten that this prey allocation strategy might affect the costs of competition among siblings.

6.7.2   Nestling Feeding Effort

Caring for offspring has costs that may translate into reduced residual reproductive value. In other words, investing resources (e.g. nutrients, energy, time) in rearing offspring means that fewer resources can be allocated to other important functions (e.g. self-maintenance mechanisms). This may come at a cost for the parents in terms of reduced survival perspectives or capacity to invest in the next reproductive events. This cost of reproduction in iteroparous species (those with multiple reproductive events over the course of their lifetime) gives rise to a trade-off between the number of offspring generated and future reproduction investment. Kestrels seem to operate well below their presumed maximum capacity (Masman et al., 1989), supporting the idea that parental effort may be optimised in order to limit costs for the parents. However, these costs may also be paid by the offspring if parents reduce their workload in order to give priority to their self-maintenance (*parent–offspring conflict theory*).

Much experimental work has assessed the impact of offspring rearing effort on parents. One way to manipulate the parental effort is through the enlargement or reduction of the brood size. Enlargements of the brood size (i.e. increase of workload due to more nestlings to feed) slightly increased the energy expenditure (Chapter 7) of both parents compared to either parents of control broods (~3%) or parents of reduced brood sizes (~6%) in a kestrel population in the Netherlands (Deerenberg et al., 1995). Importantly, the manipulation of workload affected the apparent survival probability (product of the probabilities of true survival and of study area fidelity) of kestrels: 60% of parents raising two extra nestlings died before the end of the first winter compared to 29% of those raising either control or reduced broods (Daan et al., 1996; see also Dijkstra et al., 1990b for effects of brood size manipulation on local survival). These results suggested that the parental investment of kestrels is plastic because they can increase or decrease it on demand. Consequently, nestlings might not be expected to suffer any important costs if parents adjust their foraging activity to offspring's demands.

Similar experimental manipulations of brood size carried out in a Finnish population of kestrels (i) found weak immediate effects (observations were limited to 48 h elapsed since brood size manipulation) on hunting effort or prey delivery rate (mainly of females; Tolonen & Korpimäki, 1996) and (ii) did not find any effects of brood size manipulation on the apparent survival or on the body mass of both parents (Korpimäki & Rita, 1996). On the other hand, the body condition of nestlings was reduced in enlarged brood sizes compared to control broods, indicating that costs were probably paid by the offspring. These results would support a fixed parental strategy of kestrels, with the brood size manipulation resulting in a trade-off between offspring quality and number. Korpimäki and Rita (1996) suggested that, compared to other geographic regions, the availability of prey in Finland changes unpredictably, probably making investment in self-maintenance rather than in offspring a better strategy. Selection would favour such a strategy because the food available during the laying phase is not a good predictor of the food amount that will be available while rearing the nestlings. Indeed, a food supplementation study suggested that the production of fledglings is limited even in years when voles are abundant (Wiehn & Korpimäki, 1997).

Management practices of the Mauritius kestrel provide a further nice example of how parental effort may translate into reduced survival (Nicoll et al., 2006). These management practices consisted of harvesting eggs. In so doing, females were stimulated to lay additional eggs, which might have been costly. On the other hand, the rearing effort of parents was reduced. Parents that experienced management showed a greater improvement in survival compared to parents that did not experience any management, particularly between the ages of 1 and 2 years (Nicoll et al., 2006).

6.7.3   Sex Differences in Parenting

Although kestrels share parental duties, males and females rely on different parental strategies to boost their own fitness, raising conflicts between them (*parental conflict theory*). The male hunts for him and for his mate from pair formation to the mid-nestling stage and for offspring since they are born. Females are responsible for most of the incubation and brooding, and engage in hunting after the mid-nestling stage. In Finland, food supplementation during the nestling stage reduced the prey delivery rate and hunting effort of females, but not that of males (Wiehn & Korpimäki, 1997). Food supplementation also increased the body mass of females, but did not affect the body mass of males (Wiehn & Korpimäki, 1997). Observations of parental behaviour during the post-fledging dependence period (Chapter 9) showed that although females drastically reduce their rearing effort or even abandon the area, males keep feeding the offspring until they are self-sufficient or feed offspring more frequently than females do (e.g. Bustamante, 1994; Vergara & Fargallo, 2008b; Boileau & Bretagnolle, 2014). These results would suggest that females and males rely on flexible and fixed parental strategies, respectively.

Similar experiments carried out on American kestrels in Canada found that males and females may actually have different strategies. Conversely to females, males responded promptly to manipulations of either brood size (Dawson & Bortolotti, 2003; see also Gard & Bird, 1990) or duration of the nestling-rearing period (Figure 6.6; Dawson & Bortolotti, 2008). Thus, at least in this Canadian population of American kestrels, males appeared to have a more plastic strategy than that of females. However, it is not always the case. Prior work found that female American kestrels reduced their provisioning rates when food-supplemented and had higher return rates than control females (Dawson & Bortolotti, 2002). Male kestrels, whose nests were

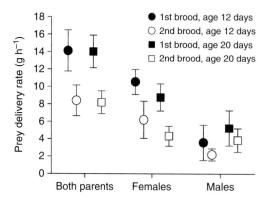

Figure 6.6 Biomass delivered per hour (± standard error) by American kestrels to nests when young were 12 and 20 days of age. Data are for natural first broods and experimentally provided second broods ($n = 11$ nest boxes). Reprinted from Dawson and Bortolotti (2008) with permission from the American Ornithological Society.

supplemented, also responded to extra food by reducing provisioning, but to a much lesser extent than their mates. Moreover, the return rates of supplemented males were similar to those of control males (Dawson & Bortolotti, 2002).

6.8   Parental Effort in Nest Defence

Defence of the nest is an important component of parental investment. Nest defence is expected to increase with brood value (e.g. number of nestlings, chances of nestlings to fledge). Males and females might also invest differently in nest defence because of potential conflicts due to different fitness pay-offs. Work on kestrels in Finland did not find much support for most of these expectations (Tolonen & Korpimäki, 1995), which were mainly derived from studies on passerine birds. Tolonen and Korpimäki (1995) found that nest defence performed by both mates against a known predator of kestrels (pine martin *Martes martes*) in the study area was generally higher in years with low availability of prey. Tolonen and Korpimäki (1995) also found that (i) early-breeder males did nest defence more than late-breeder males, (ii) the females that defended more during the incubation period were those that laid more eggs and (iii) the female effort in nest defence during the nestling period was weakly associated with the number of offspring.

In Spain, Carrillo and Aparicio (2001) also found that females' nest defence effort against humans (a known predator in the study area) was not associated with brood size. Similarly, Carrillo and González-Dávila (2013) found that the level of aggression in intrasexual encounters was not correlated with clutch size or with brood size in both sexes. In contrast, Wiklund and Village (1992) found that aggression against female decoys was higher in females that had a higher number of offspring. This might indicate that the nest defence effort of females depends on the type of intruder into the territory.

During the last week of incubation, Csermely et al. (2006) found that females attacked a decoy of a known predator (hooded crow) more often than that of a potential unknown predator (raven). However, the response of mates to the same intruder also appears to differ between studies. Tolonen and Korpimäki (1995) found a positive and significant correlation between the nest defence effort of mates; however, correlations were generally low. Carrillo and Aparicio (2001) also found a positive, but low, correlation between the intensity of the defence behaviour of both mates during the incubation and the first part (15 days) of the nestling-rearing period. However, such correlation disappeared during the second part of the nestling-rearing period. Carrillo and González-Dávila (2013) found low fluctuation in nest defence behaviour during the breeding season, and similar levels of aggression in both members of the pair towards male or female intruders. Moreover, the nest defence effort was not related to the availability of prey. These results would suggest that kestrels might have defended the nest site rather than the food resources or the offspring.

6.9   Fledging Success

The proximate and ultimate causes of the timing of nest departure are not fully understood. Reduced food provisioning by parents, sibling competition or achievement of the right body size to fledge are some of the ultimate factors that appear to stimulate the young to leave the nest through changes of hormonal status (Chapter 7). Common kestrels fledge when they are around one month old (27–32 days of age). Compared to the common kestrel, American kestrels fledge at a similar age (25–32 days of age; Palmer, 1988; Heath, 1997), Madagascar kestrels fledge earlier (at 23–24 days of age; Rene de Roland, 2005b), Australian kestrels fledge slightly later (31–35 days of age; Olsen & Olsen, 1980) and Seychelles kestrels (at 35–42 days of age; Watson, 1992), Mauritius kestrels (at 38–39 days of age; Cade & Digby, 1982) and lesser kestrels (at 36–40 days of age; Bustamante & Negro, 1994) fledge later. However, information about the age at fledging in some tropical kestrel species is limited to a small number of areas and nests. After fledging, the young stay close to the nest for some days (e.g. perching on a tree near the nest) and can even be seen to return frequently to the nest in the first few days after they fledged (Chapter 9).

Conversely to the predictions of the brood reduction hypothesis (Section 6.6), an experimental manipulation of both hatching asynchrony and food availability in a Finnish population of kestrels found that synchronous hatching had more short-term advantages (larger body mass of nestlings, higher fledging success) than asynchronous hatching during conditions of both food scarcity and food abundance (Wiehn et al., 2000). Moreover, the body condition of nestlings from synchronous broods did not improve after the death of a nestling (usually the smallest). Curiously, asynchronous hatching was more common in the year with more abundance of prey. When food is more abundant, kestrels also lay more eggs, which might explain the higher occurrence of asynchronous hatching (Wiebe et al., 1998a). In their experiment, Wiehn et al. (2000) did not manipulate the clutch size. Thus, the benefits of hatching asynchrony might only emerge in those years when females lay large clutches because food is abundant during the laying phase, but it crashes later in the season while nestling rearing (Wiehn et al., 2000). This hypothesis is currently untested.

Although last- and first-hatched nestlings might not differ in body condition or other traits at fledging, this does not necessarily mean that their survival or reproductive perspectives will be similar. For example, Massemin et al. (2002) found that last-hatched nestlings had a similar body size and body condition to that of first-hatched nestlings because they made compensatory growth. This compensatory growth was achieved by accelerating the growth rate, a phenomenon that is also known as catch-up growth (Metcalfe & Monaghan, 2001). Although catch-up growth allows an individual to achieve the body size that is supposed to be normal for its life stage in the short-term, such as an adequate size for fledging, it can carry costs that emerge later in life, such as a potential reduction in longevity (Metcalfe & Monaghan, 2001). Using a long-term and individual-based data set of common kestrels in Spain, Martínez-Padilla et al. (2017a) found that first-hatched siblings within a brood had a higher probability of recruitment into the population and a higher lifetime reproductive success than their later-hatched siblings (born on average 36.9 h later). This pattern was similar between males and females. The highest reproductive success of first-hatched siblings was independent from (i) their body condition or that of their mates at the time of breeding and (ii) their lifespan. Moreover, any bias due to different dispersal according to hatching order was found to be negligible (Martínez-Padilla et al., 2017a).

Conversely to the common kestrel, asynchronous broods of American kestrel had higher fledging success than synchronous broods when food was scarce as predicted by the brood reduction hypothesis (Wiebe & Bortolotti, 1995), even if parents made more visits to synchronous nests than to asynchronous ones (Wiebe & Bortolotti, 1994). Moreover, while the mortality of nestlings occurred at an advanced stage of growth period in common kestrels (Wiebe et al., 2000), it occurred early in life in American kestrels as predicted by the brood reduction hypothesis (Wiebe & Bortolotti, 1995). This early mortality might have enabled parents to save energy and nutrients that would have been lost if they were allocated to nestlings with low survival perspectives (Wiebe & Bortolotti, 1995). The reasons for such differences between common and American kestrels are currently unknown.

A reason for the inconsistencies about the effect of hatching asynchrony on the survival of nestlings may lie with the apparent capacity of females to modify primary (at fertilisation) or secondary (after hatching) sex ratios. This is important in light of the *sex allocation theory*, which postulates that resources are strategically divided between male and female offspring in order to maximise fitness. In all kestrel species, females are larger than males, thus the cost of rearing and the relative fitness outcomes are predicted to differ between sexes. Under this scenario, the sex ratio should depart from parity. For example, kestrels would generate more sons when there is less food because sons would be cheaper to rear than daughters because of their smaller body size. In Finland, Korpimäki et al. (2000) found that kestrels produced more males than females in the breeding season when vole densities were constantly low, while similar numbers of males and females were generated in the years when vole densities either increased or peaked and declined later in the season. Moreover, the body condition indices of both parents were lower in those broods where there were more female than male nestlings. Similarly, the proportion of male American kestrels at hatching increased as the food supply declined, and both male and female parents in poor physical condition were more likely to have male-biased broods than those in good condition (Wiebe & Bortolotti, 1992). The mortality of embryos and young did not appear to be linked to sex ratios, suggesting that biased sex ratios might result mainly from the female's control rather than from sex-biased mortality.

In the Netherlands, Dijkstra et al. (1990a) found that the proportion of males in broods declined during the breeding season, which was suggested to occur because males born early in the season have higher likelihood of breeding as yearlings, while there was no effect of birth date on the age of first breeding for females. Similar results were found in a population of American kestrel with migratory habits, but not in one that was resident all year round (Smallwood & Smallwood, 1998). Dijkstra et al. (1990a) also provided some preliminary data that suggested an active control of the female on the sex ratio with potential adaptive meaning. Last-born nestlings had lower survival perspectives than their siblings. The proportion of males was actually higher in the first three eggs laid early in the breeding season, while such a proportion declined with laying date (Dijkstra et al., 1990a). Thus, females might assign the sex with lower survival or reproductive perspectives to the later eggs in a follicle hierarchy (Dijkstra et al., 1990a). Similar seasonal declines in the number of males generated were also found in lesser kestrels in Spain (Tella et al., 1996a) or in American kestrels (Smallwood & Smallwood, 1998; Griggio et al., 2002). However, work on common kestrels in Finland found that the proportion of males in broods declined with season in only one of the three study years, while it increased in the other two years (Korpimäki et al., 2000). Further work in Finland confirmed that biased brood sex ratios are uncommon (Laaksonen et al., 2004a). In Finland, predictability of prey availability is lower than in other regions; this would make adjustment of brood sex ratio to local food availability a gamble (Laaksonen et al., 2004a).

Although it is still unclear whether variation in sex ratio is adaptive, several lines of evidence supported the hypothesis that male and female nestlings differ in their competitive abilities. For example, Fargallo et al. (2002) found that female nestlings were in better body condition than their male siblings in broods that were experimentally enlarged. Fargallo et al. (2003) found that female nestlings obtained more food from parents than their male siblings, but only in nests in which parents provided the nestlings with prey items small enough to be swallowed whole (without dismembering) by the nestlings. This result was independent from begging because male and female siblings did not differ in begging calls. Rather, female nestlings were always closer to the parent, suggesting a competitive advantage due to their larger body size than that of male nestlings (Fargallo et al., 2003). When all siblings in a brood were females, however, this size advantage in mixed-sex broods came at a cost for females in terms of reduced body condition at fledging (Laaksonen et al., 2004b).

The success of reproduction is also dependent on the weather conditions. Both the amount and timing of rainfall may affect reproductive success. Heavy rainfall may increase chilling of nestlings (Newton, 1979, 1998) or decrease parents' hunting efficiency, thus increasing the likelihood of starvation or loss of energy reserves of nestlings (Dawson & Bortolotti, 2000; Redpath et al., 2002; McDonald et al., 2004). In Poland, while the density of territorial kestrels was higher after mild winters (higher temperature, less days of snow cover), fledging success was only higher in warmer springs (Kostrzewa & Kostrzewa, 1990, 1991). In central Italy, fledging success was moderately higher in warmer and rainier springs in kestrels breeding in nest boxes (Costantini et al., 2010a). At population level, the fledging success was not associated with the body condition of nestlings (Costantini et al., 2009b), suggesting that factors other than food availability might have been important in determining the survival of the nestlings until fledging. Conversely, Kreiderits et al. (2016) found that fledging success was higher in warmer and drier seasons. One reason for these differences might lie with the nest type. The studies of Kostrzewa and Kostrzewa (1990, 1991) and Costantini et al. (2010a) were carried out on kestrels breeding in artificial nest boxes, which offer protection against rainfall. In contrast, Kreiderits et al. (2016) collected data from kestrels breeding in building cavities, which also offer good shelter, and in crow nests, which do not do so.

Positive effects of moderate rainfall on the survival of nestlings until fledging were also found in other kestrel species, and these effects came through an increased availability of prey. In Seychelles kestrels, higher breeding success was recorded in wetter years, which coincided with higher abundance of prey (green geckos belonging to the genus *Phelsuma*; Watson, 1992). Similarly, in dry springs, lesser kestrels generated fewer offspring and in poorer body condition compared to those generated in wet springs (Rodríguez & Bustamante, 2003).

6.10   Divorce Behaviour

Social monogamy is a widespread phenomenon in vertebrates, characterised by mate retention between consecutive breeding seasons. Mate retention is intrinsically dependent on pair coordination, which drives the success of parental care in biparental species. Indeed, a lack of pair coordination can result in suboptimal parental care and consequently reduced breeding success. Taken to the extreme, this can lead to a failed breeding attempt. Divorce, i.e. separation from the mate and subsequent re-mate with a new partner, may be an adaptive response to an extreme lack of pair coordination (i.e. failed breeding attempt) and is expected to increase the lifetime reproductive success. Although in socially monogamous birds divorce is widespread (Culina et al., 2015), very little is known about divorce behaviour in kestrels.

Village (1990) observed that both male and female common kestrels were more likely to breed with the same partner when they succeeded in their previous breeding attempt. Moreover, he observed that females were more likely to divorce than males because they abandoned the nest and moved to a new area. Vasko et al. (2011) also found that females were more likely to divorce; 82% of females that returned to breed in the study area, and whose prior mate was still alive, mated with a different male. In contrast, Steenhof and Peterson (2009) observed in American kestrels that (i) mate fidelity was related to nest box fidelity but not to prior nesting success or years of nesting experience and (ii) individuals that switched mates and nest boxes did not improve or decrease their subsequent nesting success.

6.11 Conclusions

We have learned much about the reproductive biology of kestrels. Males and females have evolved different parental strategies that need to be coordinated in order to increase the probabilities of reproductive success and to re-mate together the next breeding season. Males have a main role as food providers, while females have a strong control on the reproductive outcomes through the egg generation and incubation behaviour. These strategies appear to vary among kestrel populations, indicating that the magnitude of selective pressures and the way they operate on reproductive strategies may vary in space.

There are a number of potentially unexplored factors that call for cautiousness in the interpretation and generalisation of results. For example, it is unclear the degree to which weather conditions affect reproductive decisions and reproductive success independently from their indirect effect on the availability of prey. The interpretation of food supplementation experiments is also not straightforward, because supplemented parents might reduce foraging activity or use food supply as a clue to predict the availability of prey during the nestling period. Northern populations of kestrels are migratory and breed in environments with marked within- and between-year fluctuations in food abundance, whereas southern populations are almost sedentary, experience lower fluctuations in the availability of prey and have a richer diet in terms of prey types. Because of these differences, kestrels have probably evolved different behavioural (and physiological) strategies in response to various selective pressures because they are on different time schedules due to different lengths of the breeding season. We also know very little about the costs of various reproductive behaviours. For example, incubation costs might be lower for females breeding in warmer environments as suggested for the Australian kestrel (Olsen & Baker, 2001). Also, the patterns of embryo development might vary across various thermal environments owing to different intensity of abiotic stressors (e.g. egg water loss; Bird & Laguë, 1982a; Olsen & Olsen, 1987). Thus, more attention should be given to understand the extent to which reproductive strategies are also shaped by temporal and spatial environmental heterogeneity in abiotic conditions.

7 Ecological Physiology and Immunology

7.1 Chapter Summary

The quantification of physiological and immunological functions provides fundamental information on the individual state. It fosters our understanding of the costs of and constraints on life-history strategies. Research in this area on kestrels has mainly focused on immunity, energetics, hormones and antioxidants. This chapter discusses the factors that impact the immune function and describes a number of parasites and pathogens that can be detected in kestrels. It shows how the different phases of reproduction face males and females with different energetic and physiological demands. It discusses the costs associated with sibling competition and how male and female nestlings may differ in how they optimise the trade-off between growth and self-maintenance. Finally, this chapter describes the moult phase, which represents an understudied feature of kestrels' biology.

7.2 Introduction

Ecological physiology aims to describe the functions and mechanisms that govern how organisms adapt endogenously to their environments in order to answer fundamental questions: how do organisms tackle environmental changes? Why is there large diversity in physiological traits within species? Does a given function or mechanism have always the same fitness outcomes across time and contexts? Ecological physiology seeks to answer these and other questions about the short- and long-term physiological changes that occur in organisms over their lifetime, how these changes mutually interact with behavioural or life-history strategies, and which are the fitness consequences.

Immunity is another fundamental endogenous function. Ecological immunology is a discipline that aims to describe and explain the coevolution between hosts and parasites and the reasons for the large variation in immune function (Sheldon & Verhulst, 1996; Martin et al., 2011). Physiological and immunological functions are strictly intertwined and determine the individual state or phenotypic quality. This points out the importance of combining ecological physiology and ecological immunology to further our understanding of the causes and fitness consequences of individual variation in behavioural and life-history strategies (Figure 7.1).

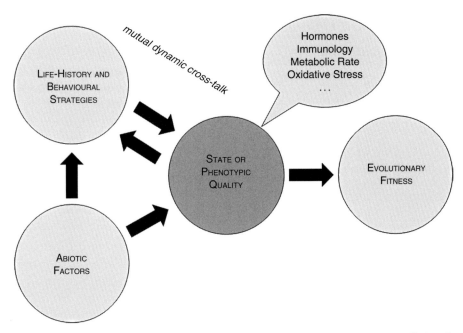

Figure 7.1 Scheme illustrating the dynamic cross-talk between individual state (or quality) and life-history or behavioural strategies and the impact on fitness. Individual state is multidimensional and is quantified combining information about different organism functions (e.g. immunity, endocrine system, oxidative stress).

7.3   Immune Function and Parasites

An organism's homeostasis is critically reliant on its immune function, which provides protection against parasites, pathogens, toxic substances and cancerous cells, and allows recovery from injuries. The immune system comprises innate and acquired components. The innate components employ constitutively produced receptors that bind to distinct molecular structures (i.e. those of microbes) and activate the host's immune cells, resulting in reactions like phagocytosis and inflammation. In contrast, the acquired immune components develop with exposure to pathogens and depend on the generation of a diverse repertoire of immune cells (e.g. T and B lymphocytes) that target specific antigenic configurations. Young birds also rely on so-called passive immunity, because mothers pass antibodies to their offspring by depositing them into the egg. Birds invest energy and nutrients to sustain their immune function, which gives rise to trade-offs with other important functions, such as growth or reproduction. Moreover, there are subtle costs associated with an immune response, such as increased molecular oxidative damage (Section 7.7.2).

In kestrels, there is a paucity of data about the organs (e.g. spleen, thymus) responsible for the synthesis of immune cells and molecules. The bursa of Fabricius is a lymphoid organ that is found only in birds. It plays a central role

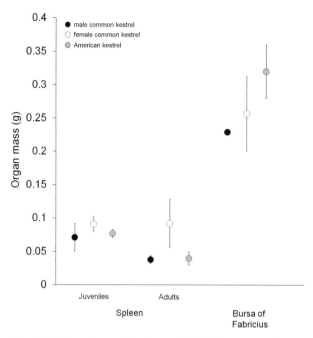

Figure 7.2 The spleen and the bursa of Fabricius are two important organs that regulate the immune function in birds. The bursa of Fabricius is only found in juveniles. Data about common kestrels are from Møller et al. (1998), data about American kestrels are from Fernie et al. (2005) and from Fallacara et al. (2011). Values are shown as mean ± standard error.

in the synthesis of B lymphocytes (which secrete antibodies) in juvenile birds and regresses before sexual maturity (Figure 7.2). Apart from a few published data on its size (Møller et al., 1998), we know almost nothing about the causes and consequences of individual variation in its function. The spleen is another important organ involved in the regulation of the immune function because it produces lymphocytes that are responsible for both the humoral antibody response and the cell-mediated immune response. A study based on a small sample size suggested that adult female kestrels might have a larger spleen size than adult male kestrels, implying better immune protection for females (Figure 7.2; Møller et al., 1998).

7.3.1   Nestlings

Animals do not appear to grow at the maximum rate (Blanckenhorn, 2000), which is peculiar given the potential benefits of reaching an increased body size quickly, including reduced predation risk in the nest or earlier time to sexual maturity, and so increased lifetime reproductive success (Dmitriew, 2011). This implies that there are constraints on growth that might work through a resource allocation trade-off with other functions, such as the immune function. Such a trade-off may be exacerbated by competition among siblings for food, which is known to increase energetic demands further.

An experimental stimulation of the T-cell-mediated immune response (thymus-dependent function) by the injection of 0.1 ml of phytohaemagglutinin (0.3 mg of PHA diluted in 0.1 ml phosphate-buffered saline) showed that female kestrels had a stronger immune response than male kestrels when food occurred in limited supply (Fargallo et al., 2002). These sexual differences were significant even when controlling for body condition or body mass, suggesting that physiological functions other than those associated with growth were likely in competition with the immune function in males (Fargallo et al., 2002).

Begging behaviour was not associated with the immune response (Fargallo et al., 2003). Female nestlings obtained more food from parents than males in those nests where parents provided the nestlings with prey small enough to be swallowed whole (without dismembering) by the nestlings (Fargallo et al., 2003). In these cases, females were also always closer than males to the parents, suggesting that they might have competitive superiority over males in food acquisition. This raises the question of whether the sex ratio of the brood may affect the young through sibling competition. Laaksonen et al. (2004b) found that female nestlings in experimentally created all-female broods had lower haematocrit (the volume of erythrocytes in blood; an index of nutritional condition) than those in mixed-sex broods, but only when the abundance of voles was low. Males from mixed-sex broods had moderately lower T-cell-mediated immunity and higher ratios of heterophiles onto lymphocytes (index of stress) than males from all-male broods. Blood stress proteins were also higher in nestlings with a low position in the brood hierarchy (Martínez-Padilla et al., 2004).

The relevance of within-brood competition in modulating the immune function was further corroborated by an experiment carried out on a Spanish population of kestrels (Martínez-Padilla & Viñuela, 2011). The immune response to the injection of 0.1 ml of a PHA solution (0.3 mg PHA in 0.1 ml phosphate-buffered saline) was stronger when brood size reduction (due to nestling mortality) occurred in asynchronous broods. In contrast, the immune response was weaker in synchronous broods where at least one nestling died before PHA injection than in synchronous broods where all nestlings fledged (Martínez-Padilla & Viñuela, 2011). Moreover, the immune response of females was stronger when all nestlings in the brood survived than when there was a brood size reduction. In contrast, the immune response of males was not related to the brood reduction. In asynchronous broods, where brood size reduction occurred, the immune response was stronger in smaller nestlings, suggesting a trade-off between growth and T-cell-mediated immunity (Martínez-Padilla & Viñuela, 2011). Although reducing growth might be a disadvantage in within-brood competition for food, this strategy might be beneficial in the post-fledging phase because it could give smaller individuals a better hunting capacity than that of larger individuals (Vergara & Fargallo, 2008b).

The T-cell-mediated immunity of nestlings is not only dependent on the posthatching conditions, but also on the conditions experienced within the egg. Martínez-Padilla (2006) found that the immune response to an injection of 0.1 ml of PHA (0.3 mg PHA in 0.1 ml phosphate-buffered saline) was weaker in those nestlings born from small eggs. Conversely, the immune response was stronger in nestlings born from large eggs laid by females that were food-supplemented before egg production. These results were not affected by variation in laying date, clutch size, brood size, or body condition of mothers (Martínez-Padilla, 2006). Importantly, the food supplementation did not affect the egg volume, indicating that the nestlings' immune function was probably affected by the egg quality (e.g. type of nutrients, antioxidants, antibodies, hormones) and not by the amount of yolk and albumen deposited in the egg (Martínez-Padilla, 2006). For example, hormones may affect the T-cell-mediated immune response. Fargallo et al. (2007b) increased the plasma concentration of testosterone in nestling kestrels using subcutaneous silastic tubing implants filled with crystalline testosterone propionate. Sixteen days after the implants were made, they tested the T-cell-mediated immune response to an injection of 0.1 ml of PHA (0.3 mg PHA in 0.1 ml phosphate-buffered saline). The immune response was significantly lower in both males and females whose testosterone level was experimentally increased and it was not linked to growth (Fargallo et al., 2007b). Although 17% of the implanted nestlings had a concentration of testosterone that was out of the natural range of control nestlings, the results of this study indicated a possible direct or indirect effect of testosterone on some traits of the immune function.

Age is also an important factor because the immune function becomes fully competent with time. A study carried out in central Italy on free-living kestrels found that older nestlings had higher counts of lymphocytes and lower counts of heterophils (Dell'Omo et al., 2009). In contrast, counts of monocytes, eosinophils and basophils were similar between younger and older nestlings. Age may also be associated with parasite burden. *Carnus hemapterus* is a small-bodied and partly black-coloured Dipteran insect. In their adult stage of life, they are blood-sucking ectoparasites of nestlings in many bird species. *C. hemapterus* is also a common ectoparasite in both common and American kestrels. Most infestations occur in nestlings within the first two weeks of life (Dawson & Bortolotti, 1997; Kal'avský & Pospíšilová, 2010; Sumasgutner et al., 2014b) and are easily observable under the wings. Although they can cause wounds, detrimental effects on nestlings were not observed (Dawson & Bortolotti, 1997; Kal'avský & Pospíšilová, 2010; Sumasgutner et al., 2014b).

Another important question is whether the variation among nestlings in immune markers predicts differences in short-term survival prospects (i.e. probability to fledge). Parejo and Silva (2009a) found that nestling kestrels with higher natural antibodies and lower heterophil counts had higher survival probability until fledging, respectively. Experimental studies are very much needed to test the association between immune function and survival.

7.3.2   Adults

The reduced capacity of adults to produce an immune response following exposure to a given pathogen can be a consequence of a large investment of resources in reproduction. Food availability may be a link between immune response and reproduction because both immune and reproductive functions require energy and nutrients.

Moreover, the availability of food may in turn affect the foraging effort, generating additional energetic demands for the individual.

In Finland, Korpimäki et al. (1995) found that *Haemoproteus tinnunculi* and *H. brachiatus* were the most frequent blood parasites in kestrels, while *Plasmodium circumflexum*, *Trypanosoma avium* and *Leucocytozoon toddi* occurred at low frequencies. Wiehn and Korpimäki (1998) found that the number of male kestrels infected with blood parasites (*Haemoproteus* and *Trypanosoma*) was larger in years with low availability of prey irrespective of whether kestrels were supplemented with food. The number of infected females was also larger in years with low abundance of prey, but, conversely to males, supplementation of food reduced the number of infected females. Wiehn and Korpimäki (1998) also found that (i) a greater hunting effort was associated with elevated parasitaemia in females, but not in males, in low food years only and (ii) a smaller hunting effort of males increased the likelihood of parasitaemia of their mates. Given that females may increase their parental effort to compensate for low contributions of their mates, it might be that such increased effort had detrimental effects on their immune function. Although this experiment did not enable to separate the effects of the several factors involved (e.g. physical effort in hunting flight, availability of nutrients and energy, probability of encountering parasites) on the parasitaemia, it points out the importance of environmental conditions in mediating the trade-off between reproduction and immune function. Correlative data collected from kestrels in Spain actually found that the probability of parasitism by *Haemoproteus* was lower in adults with higher innate humoral immunity (Parejo & Silva, 2009a), indicating that this particular component of the immune function might play an important role in controlling infection by blood parasites.

In a further study, Wiehn et al. (1999) found that the prevalence of *Trypanosoma* in males (main food provider of young) was higher in experimentally enlarged broods, and that the difference in *Trypanosoma* prevalence between reduced and enlarged broods increased with decreasing food supply. In females, the *Trypanosoma* prevalence was higher in experimentally enlarged than reduced broods, but this difference in prevalence was apparent in the year of relatively high vole densities only (Wiehn et al., 1999). Manipulation of brood size did not have clear effects on *Haemoproteus* infection in both males and females.

7.3.3   Overlooked Parasites and Pathogens

Research on the ecological relevance of parasites and pathogens in kestrels has been restricted to a few taxonomic groups. This limits our understanding of the role of single parasites or co-infections with multiple parasites in regulating population dynamics or affecting the expression of key life-history traits. Work carried out on birds housed in rehabilitation centres and veterinary screenings of carcasses of free-living birds have identified a large number of parasites and pathogens that would deserve more attention.

Babesiosis is a malaria-like parasitic disease caused by infection of erythrocytes with *Babesia*, which is transmitted by ticks. Several studies found that common kestrels infected with *Babesia* eventually died (Mohammed, 1958; Muñoz et al., 1999; Peirce, 2000). Raida and Jaensch (2000) described a protozoan infection

(*Leucocytozoon*-like infection) causing severe endarteritis, meningoencephalomyelitis and pectenitis in free-living Australian kestrels. Lemus et al. (2010) found that elevated levels of antibodies in both nestling common and lesser kestrels were associated with *Chlamydia* infections, a genus of bacteria that can have detrimental or even fatal consequences for the host. Lemus et al. (2010) also observed that in the same areas where kestrels were found infected, a *Chlamydia* outbreak was observed in both sheep and insects, suggesting a cross *Chlamydia* infection among livestock, wild insects and wild birds. Interestingly, *Chlamydia* infection was more common in lesser than common kestrels, which might have been because lesser kestrels feed primarily on insects. This reminds us of the importance of the life cycle of parasites and of host–parasite interaction, a topic that has received little attention in kestrels (Hoogenboom & Dijkstra, 1987).

Another pathogen that might deserve more attention is *Caryospora*, a genus of parasitic protozoa responsible for a variety of both acute and chronic diseases that cause significant mortality and morbidity in wild birds of prey, especially in *Falco* species (Forbes & Simpson, 1997; Forbes & Fox, 2005). Avian cholera (caused by the bacterium *Pasteurella multocida*) and tuberculosis (caused by the *Mycobacterium avium* complex) are infectious diseases that occur in kestrels (Steinhagen & Schellhaas, 1968; Smit et al., 1987, 17 infected kestrels with *Mycobacterium* out of 450 examined) and that would be relevant to study in more detail given their lethality. We also know little about the impact of endoparasites, such as acuaroid nematodes (Acosta et al., 2010), on the health of kestrels.

Viruses are another very important group of pathogens that has received surprisingly little attention so far in ecological and behavioural sciences. Herpesviruses is a large family of DNA viruses that cause latent infections and certain fatal diseases in many bird species. Diseases are characterised by depression of normal activity, respiratory distress, extremity paralysis, head-shaking and severe tissue necrosis (e.g. haemorrhagic and necrotic lesions to liver, spleen or bone-marrow; Thomas et al., 2007). Herpesvirus strains in falcons were first isolated from the prairie falcon (*Falco mexicanus*) and, later, from many other falcon species, including common kestrels (Falconid herpesvirus-1 or inclusion body disease of falcons; Ward et al., 1971; Greenwood & Cooper, 1982; Reuter et al., 2016) and American kestrels (Potgieter et al., 1979). Experimental infection with a herpesvirus strain causes a disease with a short incubation time that leads kestrels to death within around 6 days from virus inoculation (Graham et al., 1975).

Other important viruses that were isolated from free-living common kestrels are West Nile virus (Work et al., 1955), poxvirus (Kitzing, 1980) and polyomavirus (Johne & Müller, 1998). Although infections with these viruses may be fatal for kestrels (Thomas et al., 2007; Hall et al., 2009), there has not been any systematic study that addressed the causes and consequences of individual infections or viral outbreaks at population level.

The impact of pathogens on the organism is also dependent on the ecology of the host species. Kestrel populations differ in many aspects that can favour or limit the spread of pathogens. For example, in Europe, northern populations are migratory, which might make them carriers of pathogens. Isolation of island kestrels is also important. Hille et al. (2007) found that the prevalence of blood parasites in the two subspecies of common kestrel endemic to the Cape Verde islands was much lower than that recorded in kestrels from the European mainland. Abundance of parasites in kestrel populations may also depend on the population dynamics of prey that are intermediate hosts of the parasite. This is well illustrated by the lifecycle of the protozoan *Sarcocystis cernae*. This parasite undergoes asexual multiplication within the vole (intermediate host). The infected vole is preyed on by the kestrel (definitive host), within which it reproduces sexually (Hoogenboom & Dijkstra, 1987).

7.4   Endocrine System

7.4.1   Corticosterone

Glucocorticoids are a class of steroid hormones that regulate many physiological functions. Production of glucocorticoids increases when the organism is being exposed to stressful conditions. These hormones, through a cascade mechanism, activate the physiological stress response (Sapolsky et al., 2000; Romero, 2004). In birds, the production of corticosterone (main glucocorticoid in this taxon) is regulated by the hypothalamic-pituitary-adrenal axis (Sapolsky et al., 2000; Romero, 2004). Corticosterone regulates many organism functions, such as hepatic gluconeogenesis, some aspects of the immune function, or glucose metabolism (e.g. Munck et al., 1984; Romero et al., 2009). The production of corticosterone is increased within 3 minutes (stress response paradigm) of being exposed to a stressful situation, such as exposure to a predator. Chronic high levels of corticosterone production are responsible for many physiological changes and associated behavioural responses, such as inhibition of reproductive activity or growth, increase in anxiety, and changes in foraging and feeding rate (Sapolsky et al., 2000). This set of physiological and behavioural changes characterises the so-called *emergency life-history stage* (Wingfield et al., 1998). When an individual enters this stage, resources are mostly used to sustain those functions that are essential for self-maintenance and survival (McEwen & Stellar, 1993; Wingfield et al., 1998). This means that other functions, such as growth or reproduction, can be suppressed.

To assess the effects of corticosterone on the postnatal growth of nestling common kestrels, Müller et al. (2009) elevated the concentration of plasma corticosterone for 2–3 days using subcutaneous biodegradable implants that release the hormone into the organism. The plasma concentration of corticosterone induced by the implants was comparable to that caused by either nutritional or handling stress (Müller et al., 2009). The experimental increase of corticosterone caused a reduction in the growth rate of both feathers (71%) and skeletal traits (14–26%) compared to those nestlings whose corticosterone level was not increased. Treated birds also did not gain body mass as controls did, but their subcutaneous fat stores were unaffected. After corticosterone levels returned to normal, nestlings could recover their body mass and, to some degree, their tarsus length through compensatory growth (Müller et al., 2009). Thus, increased corticosterone during the developmental period may induce a selective suppression of growth, where skeletal growth is prioritised at a cost for more flexible body tissues, such as muscle or feathers (Müller et al., 2009).

Further work showed that 10-day-old kestrels already have the capacity to elevate corticosterone production in response to a stressful event (Müller et al., 2010) as previously demonstrated for nestling American kestrels (Love et al., 2003a). Moreover, it has been found that the stress-induced increase in corticosterone production is stronger in older nestlings in both common and American kestrels (Love et al., 2003a; Müller et al., 2010); this result shows that the stress-responsiveness function needs time to become fully mature and is highly conserved in kestrel species.

The lower capacity of young nestlings to elevate corticosterone levels in response to stress and the slow maturation timing of this function might be explained by the costs associated with it, such as suppression of growth (Müller et al., 2009) and of some components of the immune function (Müller et al., 2011) or increased generation of molecular oxidative damage (Section 7.7). For example, an experimental increase of corticosterone reduced the T-cell-mediated immune response to an injection of 0.05 ml of PHA (0.25 mg PHA dissolved in 0.05 ml phosphate-buffered saline) at the end of the corticosterone treatment and, to a lesser degree, 5 days after circulating corticosterone had returned to baseline concentrations (Müller et al., 2011). Thus, development of endogenous functions needs to be optimised in order to prioritise those that are more important at that specific stage of growth or can be less costly to sustain.

Increased production of corticosterone in response to an acute stress exposure is certainly important. However, individual variation in baseline corticosterone may also translate in fitness outcomes. Strasser and Heath (2013) found that as baseline corticosterone concentrations in female American kestrels gradually increased due to human disturbance, the probability of failure also significantly increased. Chronic elevation of baseline corticosterone might, therefore, have a number of downstream effects that probably result in reduced reproductive effort in order to allocate resources to self-maintenance functions. Heath (1997) also indicated a potential relationship between corticosterone levels and fledging time in nestling American kestrels, suggesting that high circulating corticosterone levels could trigger movement from the nest.

7.4.2   Melatonin

Melatonin is a hormone that regulates the circadian rhythms of several biological functions (body temperature regulation: Pang et al., 1991; metabolism: Zeman et al., 1993; antioxidant activity: Reiter et al., 1999; Yilmaz & Yilmaz, 2006). Little is known about its natural variation and function in kestrels and, more generally, in birds. Dell'Omo et al. (2009) found that the serum concentration of melatonin was lower in nestling kestrels that were almost ready to fledge compared with younger nestlings (Figure 7.3). Given its multiple regulatory functions, it might be that younger nestlings

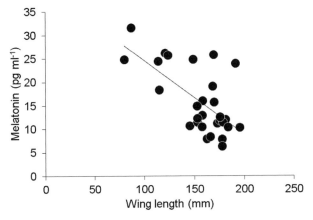

Figure 7.3 Older nestlings (wing length as a proxy of age) show lower concentrations of melatonin in serum. Reprinted with slight modifications from Dell'Omo et al. (2009) with permission of Elsevier.

have a greater need of melatonin in order to regulate different vital functions at the beginning of their maturation phase.

7.4.3   Sexual and Reproductive Hormones

Meijer and Schwabl (1989) were the first to describe the hormonal patterns in common kestrels across the year in the Netherlands. Plasma levels of the luteinising hormone (LH) and total androgens (testosterone and 5α-dihydrotestosterone) were measured in breeding and non-breeding kestrels, both in the field and in captivity. LH stimulates ovulation in females and sexual hormone production in both males and females. LH was low in free-living males in winter, but rose in February–March when other males arrived at the breeding grounds and most territories were established. LH remained at high levels throughout incubation and declined to basal levels during the nestling feeding period (Meijer & Schwabl, 1989). Temporal changes of LH in captive males were similar to those recorded in free-living males. Seasonal changes in LH plasma levels in males of non-breeding and breeding pairs were similar, except during December, when males of those pairs that subsequently bred had significantly higher LH concentrations. In females, LH rose during courtship and reached a maximum during egg-laying (Meijer & Schwabl, 1989). The plasma concentration of LH dropped during incubation and declined further to winter levels during the last stages of the nestling-rearing phase. In captive females, as in males, LH levels were already somewhat elevated during winter compared with free-living birds. Moreover, LH was higher in laying than non-laying females, as well as in those females that engaged in courtship behaviour compared to those that did not (Meijer & Schwabl, 1989).

Androgens were low in free-living males during winter and reached peak concentrations during courtship, dropping to basal levels during incubation. In contrast, the rise in androgens during the courtship phase was completely absent in captive males

(Meijer & Schwabl, 1989). Androgen levels of both free-living and captive non-breeding males were generally elevated during spring. However, their concentrations were much higher in free-living than in captive males (Meijer & Schwabl, 1989).

A study on both free-living and captive common kestrels during the breeding season in central Italy analysed in more detail the temporal changes of the two androgens testosterone and 5α-dihydrotestosterone and of the estradiol, which is the major female sexual hormone (Casagrande et al., 2011). In breeding free-living kestrels, androgens decreased from the mating to the nestling-rearing period as previously observed by Meijer and Schwabl (1989). This decrease was also evident in non-breeding captive kestrels for 5α-dihydrotestosterone, but not for testosterone. In free-living birds, plasma concentrations of both androgen types were higher in males than females during mating, while they did not differ between sexes during the nestling-rearing phase (Casagrande et al., 2011). Conversely to Meijer and Schwabl (1989), testosterone was lower in breeding free-living males during the nestling-rearing period than in non-breeding captive males. In free-living kestrels, estradiol was higher in females than males during the mating but not during the nestling-rearing phase (Casagrande et al., 2011). The plasma concentration of estradiol was always below the detection limit of the laboratory assay in captive non-breeding birds, indicating that it occurred in blood at very low concentrations.

In contrast to androgens and estrogens, other hormones that regulate reproductive behaviour have received comparatively much less attention. For example, prolactin is an important hormone involved in the regulation of parental care (Chapter 6). Meijer et al. (1990) found that the concentration of prolactin in plasma of captive kestrel females increased significantly from winter to the nestling-rearing phase. A significant role of prolactin in the regulation of incubation behaviour was later demonstrated experimentally in the American kestrel (Sockman et al., 2000).

7.5   Energy Expenditure

In the 1980s, scientists at the University of Groningen carried out a number of studies on the ecological energetics of common kestrels. All ecological processes that are crucial for key life-history traits like growth, survival and reproductive fitness demand and compete for energy. Work done on a small number of kestrels found that variation in basal metabolic rate (metabolic rate measured in an animal in postabsorptive state, at rest and at temperatures within the thermoneutral zone) was primarily explained by lean mass of the two metabolically highly active tissues, the heart and the kidney (Daan et al., 1989b), indicating that energy demands also vary across tissues.

7.5.1   Nestlings

Nestlings face important energetic demands to sustain their growth and development (see Figure 6.1). Pen (2000) found that growth curves of nestlings differed in terms of asymptotic body mass and rate among populations from different latitudes (i.e. Finland,

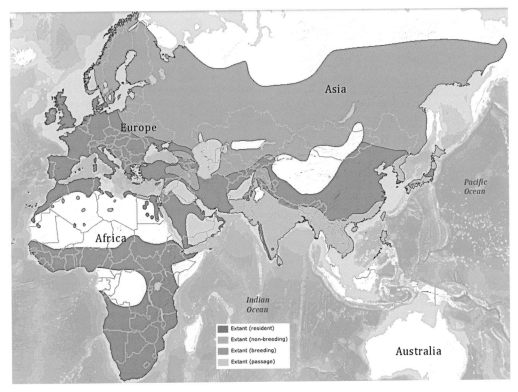

Figure 1.5 Distribution map of the common kestrel (*Falco tinnunculus*). From BirdLife International and Handbook of the Birds of the World (2017). Reproduced with permission from BirdLife International.

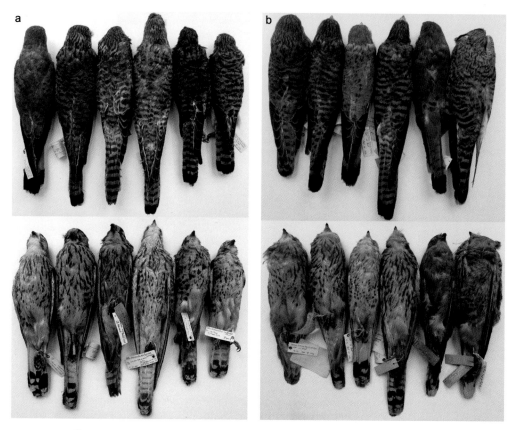

Figure 1.6 Comparison of subspecies of *Falco tinnunculus* and of other kestrel species. Panel A from left to right: *F. t. tinnunculus* (male, C.G. 1967 N. 1709, MNHN); *F. t. tinnunculus* (female, C.G. 1912 N. 560, MNHN); *F. t. interstinctus* (female, C.G. 2003 N. 159, MNHN); *F. t. rufescenes* (female, C.G. 1977 N. 347, MNHN); *F. t. alexandri* (female, C.G. 1966 N. 904, MNHN); *F. t. neglectus* (female, C.G. 1967 N. 1756, MNHN). Panel B from left to right: *F. t. dacotiae* (female, C.G. 1965 N. 1492, MNHN); *F. t. canariensis* (female, C.G. 1911 N. 882, MNHN); *F. t. canariensis* (male, C.G. 1965 N. 1484, MNHN); *F. moluccensis* (female, C.G. 1882 N. 152, MNHN); *F. rupicolus* (male, C.G. 2018 N. 503, MNHN); *F. r. rupicoloides* (female, C.G. 2003 N. 165, MNHN). Specimens are from the collection of the Muséum National d'Histoire Naturelle (MNHN; Paris, France). The MNHN gives access to the collections in the framework of the RECOLNAT national Research Infrastructure. Photographs by David Costantini.

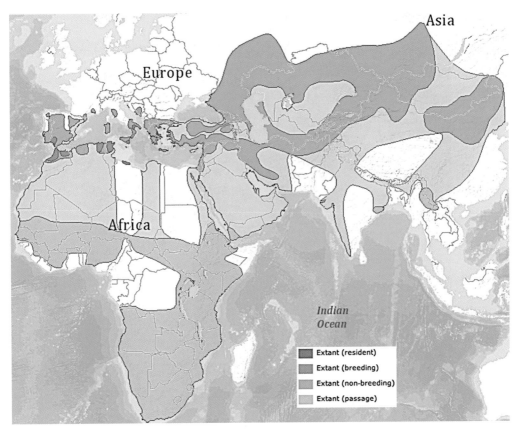

Europe

Asia

Africa

Indian
Ocean

Extant (resident)
Extant (breeding)
Extant (non-breeding)
Extant (passage)

Figure 1.8 Distribution map of the lesser kestrel (*F. naumanni*). From BirdLife International and Handbook of the Birds of the World (2017). Reproduced with permission from BirdLife International.

Figure 2.2 A male kestrel is decapitating an Italian three-toed skink (*Chalcides chalcides*). Photograph by Gianluca Damiani.

Figure 2.4 Western green lizards are an important prey species for kestrels in central Italy (Costantini et al., 2005, 2007c). Photograph by Gianluca Damiani.

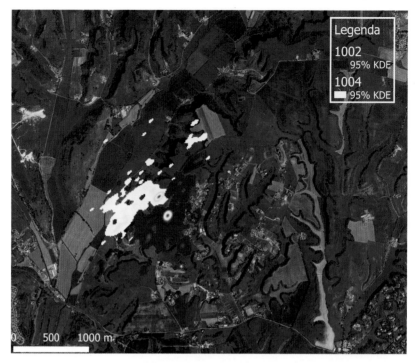

Figure 3.3 Home ranges of two female kestrels breeding close to each other (around 300 m) estimated by the fixed kernel method.

Figure 3.4 Home ranges of a kestrel pair estimated by the fixed kernel method.

Figure 5.1 Kestrels are medium-size raptors approximately 32–38 cm long, with a wingspan of around 68–78 cm. Females are approximately 20% larger than males, but overlap occurs between the smaller females and the larger males. This image illustrates a large within-sex variation in body size (the first three kestrels from the left are males). Specimens are from the collection of the Muséum National d'Histoire Naturelle (MNHN; Paris, France). The MNHN gives access to the collections in the framework of the RECOLNAT national Research Infrastructure. Photographs by David Costantini.

Figure 5.3 A male is passing a prey to his mate. In kestrels, the male provides food to the female from before the start of egg laying until nestlings are in the middle of their growing period. Photograph by Gianluca Damiani.

the Netherlands and Spain). The change in daily energy expenditure with age was mostly explained by individual body mass, while sex, population and age were poor predictors. These results suggested that nestlings of similar size would face with similar energetic demands irrespective of their sex, age and location where they are grown (Pen, 2000).

Hatching asynchrony may cause a body size hierarchy, which amplifies variation among siblings in the growth rate and in body size. Massemin et al. (2002) found that last-hatched nestlings were smaller than their siblings at 19 days of age, but were of similar size when they were 26 days old. Thus, last-hatched nestlings increased their growth rate in order to catch up with their siblings (compensatory growth). This increased growth rate resulted in a significant increase of resting metabolism and in a reduction of body condition index (Massemin et al., 2002). Accelerated growth may have a number of detrimental long-term effects on the individual that emerge at adulthood, but these effects have never been explored in common kestrels, nor have the underlying mechanisms been identified.

Massemin et al. (2003) found that, without food supplementation, the daily energy expenditure (measured using doubly labelled water) of first-hatched nestlings (~340 kJ day^{-1}) was 35% higher than the last-hatched nestlings (~220 kJ day^{-1}) while controlling for age differences. However, when asynchronous broods were provided with a food supplement, the daily energy expenditure of first-hatched nestlings (~150 kJ day^{-1}) declined significantly and was similar between first- and last-hatched nestlings (~160 kJ day^{-1}; Massemin et al., 2003). Although the experiment was based on a small sample size, these results suggested that older nestlings might invest more in competition (physical activity, begging) in order to monopolise food resources when these occur in limited supply.

Interpretation of results about metabolic rate and energy expenditure in free-living animals is complicated by the lack of an appropriate control for factors, such as ambient temperature, that are known to affect metabolism. For example, rock kestrels may reduce energy expenditure at low ambient temperatures in winter and shift their thermoneutral zone between winter and summer (Bush et al., 2008). A comparative study of three European populations of kestrels suggested that climatic conditions might not be the main determinant of daily energy expenditure in nestlings (Pen, 2000). However, microclimatic variations within the study area might contribute to generate significant variation among broods in metabolic demands.

7.5.2   Adults

The daily metabolisable energy intake is one important aspect of the ecological energetics because it gives an indication of the energy available to the organism after correcting for that lost during processes such as food digestion (e.g. assimilation efficiency). Masman et al. (1986) estimated the daily metabolisable energy intake in Dutch kestrels across the different phases of the annual cycle (Figure 7.4). They found that although the nutrient intake (gross estimate of lipids and proteins) varied across the year, the energy content per gram of dry matter did not show significant seasonal changes. The daily metabolisable energy intake also varied among the phases of the

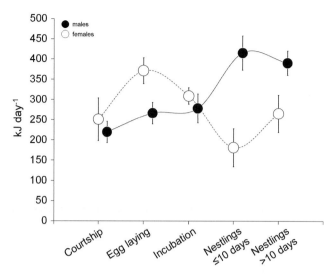

Figure 7.4 Female kestrels spend more energy during the egg-laying phase, although males face higher energetic demands during the nestling-rearing phase. Data from Masman et al. (1986).

annual cycle. In females, it was highest during egg-laying, while it was highest in males during the nestling-rearing period (Masman et al., 1986). In both sexes, the daily metabolisable energy intake was lowest during the moult phase. Masman et al. (1986) also identified a number of possible sources of variation that can affect estimates of daily metabolisable energy intake, such as time of day, weather conditions, prey energy content or amount of prey that is actually eaten.

The findings presented by Masman et al. (1986) suggested that the allocation of time to a given behaviour may be affected by the energetic demands of the behaviour itself and the energy available (Masman et al., 1988b). Strategic allocation of energy to different behaviours is particularly relevant during the demanding phase of reproduction. Large investment into reproduction may entail high energetic costs that could be paid later in life in terms of reduced future reproductive success or survival because less energy is allocated to self-maintenance functions. Analyses of energy expenditure based on several techniques actually found that males experience a large increase in energy expenditure during the whole reproductive phase, while the female experiences a peak of energy expenditure during the second part of the nestling-rearing phase, when she actively engages in foraging for her offspring (Masman et al., 1988a). Further analyses of energy expenditure showed that the egg-laying and incubation phases are also significantly energy demanding for females (Meijer et al., 1989). Moreover, early-breeder females had higher maintenance and thermoregulation costs than late-breeder females, but they also had higher daily metabolisable energy intake due to higher food intake (Meijer et al., 1989). It has also been found that energy expenditure was higher in parents rearing experimentally enlarged broods and was generally weakly linked to sex, body mass, ambient temperature, rainfall or wind speed (Deerenberg et al., 1995). Similar analyses carried out on Finnish kestrels during a comparable phase of nestling rearing (15 days old on average) found lower energy expenditure compared to the Dutch kestrels (which were bigger) and a tendency for higher energy expenditure in larger individuals and colder days (Jönsson et al., 1996).

Reproduction is not the only energy-demanding activity performed by kestrels. Moult is another phase of the annual cycle facing kestrels with important energetic demands that might increase costs of reproduction further because it partly overlaps with the reproductive phase (Section 7.8).

7.6   Body Energy Reserves

The capacity of individuals to regulate their body energy reserves (e.g. lipids, proteins) in relation to food availability and metabolic demands may have critical implications for their reproductive fitness and chances of survival. Large energetic reserves provide individuals a fail-safe for periods of food shortage or of high metabolic demands, such as harsh weather conditions or reproduction. A reduction in body reserves may also be adaptive if this reduces metabolic demands of flight or increases foraging efficiency. The adaptive meaning of body energy reserves might also vary among populations inhabiting different geographic regions, where selective pressures operate in different ways. For example, for kestrels living at higher latitudes it might be more adaptive to accumulate energy stores than for southern kestrels because at higher latitudes weather conditions can be harsher and food availability can be less predictable. However, this hypothesis is not apparently supported by a comparison of body mass of Finnish kestrels with that of kestrels in the Netherlands and Scotland (Jönsson et al., 1999) or in northern Italy (Costantini et al., 2014) (see also Table 5.1). This result might be because Finnish kestrels are migratory and start breeding within 2–3 weeks from their arrival at the breeding ground. Thus, if they loose body mass while migrating, they might not have enough time to accumulate body reserves when they are back at the breeding area.

The body condition index, i.e. body mass standardised for the skeletal size, is commonly used to quantify energy reserves in birds. Body mass may also be reliable as an index of energy reserves, particularly in longitudinal studies that record the individual's life history. In female common kestrels, body mass was significantly correlated with subcutaneous breast fat ($r_s = 0.53$, $p < 0.001$, $n = 164$; Jönsson et al., 1999). In nestling American kestrels, the body condition index was positively correlated with the plasma concentration of albumin, which is partly affected by diet (Ardia, 2006). Moreover, nestling American kestrels in better body condition had a lower plasma concentration of stress hormones (Sockman & Schwabl, 2001). In central Italy, nestlings were in better condition at population level in those years when the diet was richer in rodents, but were in lower condition when reptiles were the main prey among vertebrates (Figure 7.5).

This result might point out an important role of prey nutrient composition in determining the condition of nestlings. Apart from possible explanations, this result

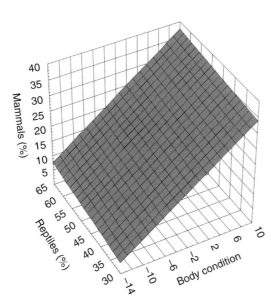

Figure 7.5 The body condition index of nestling kestrels in central Italy was higher in those years when the amount of voles in the diet was larger, although it was lower when reptiles (mostly lizards) represented the main prey at population level. The amount of birds in the diet was weakly correlated with the body condition index.

suggests that a diet richer in lizards would reduce the fitness of nestlings if the body condition index was an important determinant of the individual state. However, body condition was not associated with the fledging probability (Costantini et al., 2009b). In many bird species, body mass or body condition are good predictors of survival during wintertime and recruitment into the breeding population because body energy reserves enable individuals to cope with adverse conditions of their first winter. A long-term and individual-based study of kestrels in Spain found that first-hatched nestlings, which were in better body condition than their siblings at fledging, had higher survival probability and number of offspring produced during their lifetime than middle- or last-hatched nestlings (Martínez-Padilla et al., 2017a). Thus, long-term longitudinal studies are important to understand the adaptive meaning of the body condition index.

Within-brood competition is another important factor that generates variation in body condition among siblings. If environmental conditions get worse with season, the effect of the within-brood competition might become stronger in larger and later broods. In central Italy, the average body condition of a brood decreased over the season in broods of four, five or six nestlings, but not in those of three nestlings (Figure 7.6), and was generally higher in larger than smaller broods (Costantini et al., 2009b). Thus, only in very small broods, where sibling competition is lower, a seasonal effect on body condition was not detectable. This might also be because broods of three nestlings are generally in worse condition as compared to other broods, while low variation in condition occurs among broods of four, five and six nestlings (Costantini et al., 2009b).

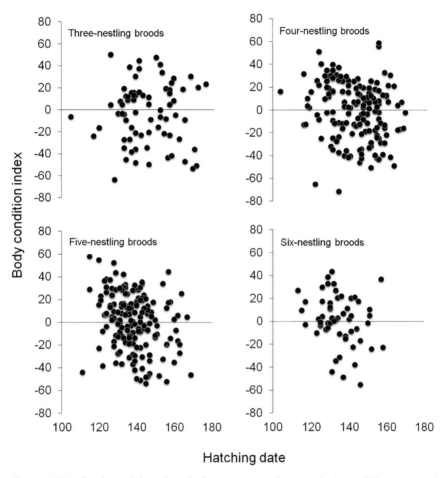

Figure 7.6 Nestling kestrels born later in the season were in worse body condition compared to those born earlier in the season in four- to six-nestling broods, but not in three-nestling broods. Data from Costantini et al. (2009b).

7.7   Antioxidants and Molecular Oxidative Damage

It is increasingly recognised that oxidative stress might be a prime physiological mediator of some key short- and long-term life-history trade-offs because the resultant tissue degradation might influence growth patterns, reproductive performance, senescence and survival (Costantini, 2014). Oxidative stress occurs when there is increased generation of molecular oxidative damage because antioxidant protection is overwhelmed by the pro-oxidant activity of free radicals and non-free radical chemicals generated by body functions (e.g. mitochondrial activity, immune response). This increased oxidative damage may have functional or fitness consequences.

7.7.1   Antioxidants

Carotenoids are the molecules that have received more attention in ecophysiological research on kestrels. Carotenoids are lipophilic pigments that animals are unable to synthesise, hence they must obtain them from the diet (Hill & McGraw, 2006a, 2006b). Kestrels display carotenoid-based colourations in the skin of lores, cere and tarsi (Chapter 5). Carotenoids occur in limited supply in the food; their concentration in plasma increases rapidly (within a week) following supplementation of carotenoids to nestlings (Figure 7.7; Casagrande et al., 2007; Costantini et al., 2007b) or adults (Costantini et al., 2007d; De Neve et al., 2008). The concentration of carotenoids in blood increases from hatching to fledgling (6.6–23.7 µg ml^{-1}; De Neve et al., 2008) and is lower in nestlings than in adults (Casagrande et al., 2007, 2011; Laaksonen et al., 2008).

Given the limited availability of carotenoids in diet and their multiple physiological functions, such as antioxidant activity, it has been hypothesised that a large allocation of carotenoids to the production of body colourations would mean that less are available to be allocated to other physiological functions. However, studies that supplemented either nestling or adult kestrels did not find any support for an important role of carotenoids as antioxidants. For example, the oral administration of about 4 mg of carotenoids (lutein and zeaxanthin) to nestlings every other day for a total of five administrations increased significantly the serum concentration of carotenoids, but did not affect a marker of serum oxidative damage or a marker of serum non-enzymatic antioxidant capacity (Costantini et al., 2007b). Thus, during the demanding period of growth and development, carotenoids appeared to play a negligible role in the protection against oxidative stress in blood.

In a study on captive adult kestrels, a prolonged supplementation of carotenoids (8 mg day^{-1} for 28 days) increased the amount of oxidative damage in serum, indicating that carotenoids may have toxic effects at high concentrations (Costantini et al., 2007d). The baseline serum concentration of carotenoids in kestrels that were not supplemented (95% confidence interval [CI]: 34.96–46.48 µg ml^{-1}) was within the range of those recorded in free-living adult kestrels during the courtship phase (95% CI: 27.81–46.80 µg ml^{-1}; Casagrande et al., 2006b). In contrast, circulating carotenoids of supplemented birds (95% CI: 2 weeks after the supplementation started was 56.56–82.14 µg ml^{-1}) were about twofold higher than the levels recorded in free-living kestrels. Although the carotenoid concentration of supplemented birds was very high, it did not appear unnatural because the carotenoid concentration may be up to 60–80 µg ml^{-1} in free-living kestrels (Casagrande et al., 2007, 2011). Carotenoids may indeed cause toxic effects at high concentrations (Palozza, 1998; van Helden et al., 2009). The potential toxic effects that carotenoids have at high concentrations might have contributed to constrain the capacity of kestrels to absorb and store carotenoids in the body.

Although evidence in favour of an antioxidant role of carotenoids in kestrels is weak at best, they may have other important roles, such as stimulation of the immune function. Costantini and Dell'Omo (2006a) found that stimulation of immune function by injection of PHA increased the serum carotenoid concentration 24 hours later,

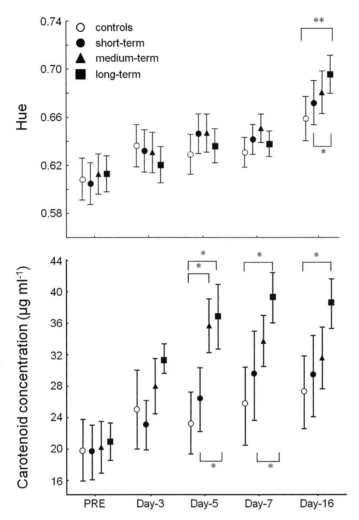

Figure 7.7 The carotenoid concentration in serum and the skin colouration of nestling kestrels show a different time course in response to carotenoid supplementation. Carotenoids are absorbed in the gut and pass relatively rapidly into the bloodstream. An increase in serum carotenoids was detectable 96 h after the start of carotenoid supplementation. In contrast, the colouration of tarsi skin (hue) showed a significant increase only after around 11 days since the start of carotenoid supplementation. At this time, the effect of supplementation on serum carotenoids was no longer evident, indicating a quick reduction in circulating carotenoids due to their storage in other tissues and use in physiological functions. Reprinted with slight modifications from Casagrande et al. (2007) with permission from Springer Nature.

suggesting that carotenoids were remobilised from other tissues to sustain an effective immune response. Similarly, the stimulation of the immune response by injection of 0.1 ml of a 2.5 mg ml^{-1} PHA solution increased the plasma concentration of carotenoids in nestling lesser kestrels (Rodríguez et al., 2014). De Neve et al. (2008) found that offspring of carotenoid-supplemented female common kestrels were infested by fewer intestinal parasite nematodes of the genus *Capillaria*, had higher lymphocyte concentrations in plasma and were less stressed (heterophile to lymphocyte ratio) than control nestlings.

7.7.2   Generators of Molecular Oxidative Damage

There is substantial evidence that oxidative stress is a cost of and constraint on growth and development (Costantini, 2014). Moreover, it is well established that the enzymatic component of the antioxidant machinery is immature in young individuals (Costantini, 2014). Costantini et al. (2006) found that younger broods had higher serum oxidative damage than older broods. Moreover, a within-individual longitudinal study showed that serum oxidative damage decreased with age in nestling kestrels (Costantini et al., 2007b). In kestrels, the growth assumes an exponential profile during their first 2 weeks of life (personal observations; Pen, 2000). Thus, the higher oxidative damage of younger nestlings might have been due to a combination of factors, such as immature enzymatic antioxidant machinery, increased metabolic rate or exposure to a higher partial oxygen pressure after hatching (Surai, 2002). In contrast to serum oxidative damage, both the serum concentration of carotenoids and the serum non-enzymatic antioxidant capacity showed inconsistent associations with age across studies (Costantini et al., 2006, 2007a; Casagrande et al., 2007). It might be that any changes in non-enzymatic antioxidants with age were masked by antioxidants that mothers deposited into the eggs or those acquired by food (Surai, 2002; Costantini, 2014). Cross-fostering experiments showed that the variation among broods in circulating non-enzymatic antioxidants was mainly explained by the environmental conditions rather than the genetic or maternal effects (Costantini & Dell'Omo, 2006b; Casagrande et al., 2009).

The immune function may be another significant source of oxidative stress. Immune cells release free radicals and non-radical pro-oxidant molecules to kill pathogens because of their cytotoxic effects (Costantini, 2014). However, the action of pro-oxidants released by immune cells is not specific, so that they may also generate oxidative damage to important biomolecules, such as DNA, lipids and proteins. Subcutaneous injection with 0.05 ml of a 1 mg ml^{-1} PHA solution increased oxidative damage and decreased non-enzymatic antioxidant protection in the serum of nestling kestrels, respectively (Costantini & Dell'Omo, 2006a). In contrast, subcutaneous injection with 0.1 ml of a 2.5 mg ml^{-1} PHA solution did not affect the non-enzymatic antioxidant protection in plasma of nestling lesser kestrels (Rodríguez et al., 2014).

Hormonal changes may also affect the blood oxidative status. Casagrande et al. (2012) found that an experimental increase of estradiol using implants filled with crystalline 17β-estradiol increased plasma oxidative damage and reduced the plasma non-enzymatic antioxidant capacity in non-breeding captive adult kestrels. In contrast, the experimental increase of 5α-dihydrotestosterone did not affect any plasma oxidative status marker. The hormonal treatment increased 5α-dihydrotestosterone until levels of about 60 pg ml^{-1}, which was within the physiological range of free-living kestrels (Casagrande et al., 2011). In contrast, the estradiol treatment generated plasma concentrations of about 400 pg ml^{-1} in females and 150 pg ml^{-1} in males, which were much higher than those recorded in free-living kestrels (close to zero in breeding males and about 10 pg ml^{-1} in females at the time of mating and close to 0 during rearing in both sexes; Casagrande et al., 2011). The effect of estradiol might have been due to these abnormal levels induced by the treatment. However, estradiol may reach very high concentrations during the egg-laying phase in females as observed in the American kestrel (Rehder et al., 1986) or in other bird species (Hunt & Wingfield, 2004). Further work is needed to clarify the links between estradiol and oxidative stress.

Corticosteroids can also affect the blood oxidative status. Costantini et al. (2008) administered corticosteroids through the diet (20 mg kg^{-1} of diet) to non-breeding captive adult kestrels over a period of 2 weeks to assess the effects of a simulated chronic stress state on the blood oxidative status. The administration of corticosterone caused a 32% increase of a marker of serum oxidative damage, but did not affect the serum non-enzymatic antioxidant capacity or the serum concentration of carotenoids.

Finally, there is some indirect evidence that the nestling-rearing effort may increase oxidative damage of males. A cross-sectional study on free-living common kestrels showed that a marker of plasma oxidative damage was higher in males during the nestling-rearing period than in males sampled prior to egg-laying (Figure 7.8; Casagrande et al., 2011). Moreover, the oxidative damage of males was similar to

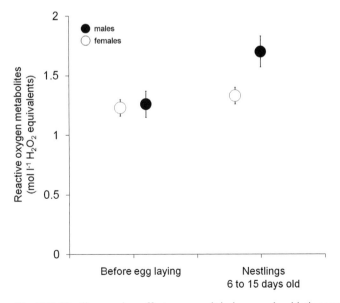

Figure 7.8 Nestling-rearing effort may result in increased oxidative stress. Plasma oxidative damage was similar between males and females before the start of egg-laying, but was significantly higher in males than in females during the nestling period. Data from Casagrande et al. (2011).

that of females prior to egg-laying, but was significantly higher in males than females during the nestling-rearing period.

7.8   Moult

Moult is defined as the periodic shedding and replacement of feathers (Campbell & Lack, 1985). Feathers are essential for body insulation and flight, but they wear out with time. Their deterioration is due to the action of parasites, physical wear or sunlight. Birds replace their old feathers with new ones by pushing the old feathers out long before the new ones are fully grown and functional. Many bird species start moulting after the end of the breeding season and the moult period may last one or several years depending on the species (Zuberogoitia et al., 2018). The post-fledging body moult of kestrels is highly variable, with some birds changing nearly all their feathers before they are one year old, whereas others changing only a few (Village, 1990). Post-fledging moult has been reported in many kestrel species, including the American kestrel (Willoughby & Cade, 1964; Balgooyen, 1976), the Mauritius kestrel (Jones, 1987), the common kestrel (Village, 1990), the Seychelles kestrel (Watson, 1992, 1993), the Australian kestrel (Olsen, 1995) and the greater kestrel (Kemp, 1999). Although little research has been done on moult in adult kestrels, the available evidence suggests that kestrels moult all their feathers once a year, replacing them in a rather fixed sequence over a period of 5–6 months (Village et al., 1980; Meijer, 1989; Young et al., 2009). Village (1990) suggested that the typical moult order for primaries in kestrels is 4–5–6–3–7–8–2–9–10–1 (Figure 7.9), which would be similar to that of other falcons. For tail feathers, a typical sequence is 1–6–2–3–4–5 (Village, 1990). Slight changes to this order can be observed; for example, the primary 7 may be lost before primary 3 (Village, 1990). The moult of primaries usually starts before that of tail feathers, and the moult of tail coverts and the rump usually starts before the underside or wing coverts, and the back and head (Village, 1990).

Masman (1986) estimated that all feathers of a common kestrel weigh about 20–25 g, which is about 10% of the body mass. Similarly, Kemp (1999) estimated that all feathers of a male greater kestrel weighed about 27 g, which was 12% of the individual body mass. The rate of feather growth calculated from five female common kestrels that were caught twice during the moult of their primaries was estimated as 25 days per feather (Village, 1990). In juveniles, the moult starts after they fledge and may last until their first summer (Village, 1990). However, the moult does not follow a strict schedule and significant variation among individuals or populations from different geographic regions occurs in the start, end and progress of moult (Village, 1990). For example, non-breeders moult faster than breeders (Meijer, 1989). In temperate regions, female kestrels start to moult during incubation, while males generally start to moult later, during the nestling-rearing period when the female shares the hunting (Village et al., 1980; Meijer,

Figure 7.9 Remiges (P = primaries; S = secondaries) and tail feathers (rectrices, RR). Primaries are numbered from the carpal joint outwards. The secondaries are numbered from the carpal joint inward (to the body) and are all those feathers attached to the ulna. Photograph by David Costantini.

1989). Although there is variation in the start of the moult, there seems to be little variation for the end of the moult period (Village et al., 1980; Young et al., 2009). The moult also seems to be faster in males than in females (Village et al., 1990; Young et al., 2009). However, a study carried out on captive kestrels found quite contrasting results (Meijer, 1989). Under natural daylight, there was no difference in the start of moult between male and female breeders, but duration of moult was longer (i.e. it was slower) in males than in females (Meijer, 1989). Under artificial daylight schedules, breeding males (generating a single clutch) started moult later than (17.5 h of daylight) or before (13 h of daylight) females, but in both cases, males moulted more slowly than did females (Meijer, 1989). Kestrels may also arrest moult (*arrested moult*) during demanding periods (Village, 1990), which complicates interpretation of moult timing.

Which proximate mechanisms regulate moult? What are the costs of moulting? Young et al. (2009) found that the moult of male kestrels started at the time of testis size regression, suggesting a significant link between reproductive

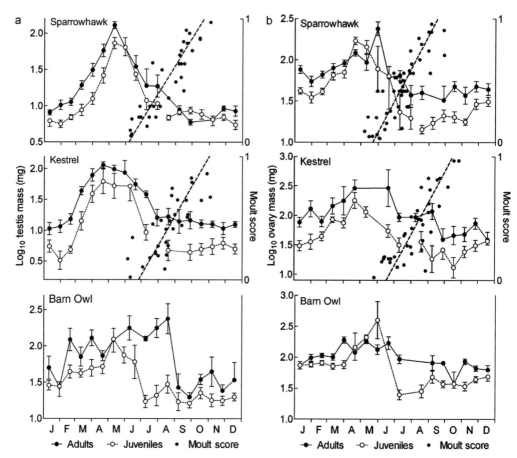

Figure 7.10 Comparison of testicular and ovarian mass between common kestrel and two other birds of prey. (a) Changes in testicular mass (mean ± standard error) during the year in adult males (solid circles and line) and juvenile males (open circles and broken line). (b) Changes in ovarian mass (mean ± standard error) during the year in adult females (solid circles and line) and juvenile females (open circles and broken line). In both panels, individual moult scores for sparrowhawks and common kestrels are shown as small solid circles and the broken line represents linear regression of day on moult score. Reprinted with slight modifications from Young et al. (2009) with permission from John Wiley and Sons.

physiology and moult. In females, there was also some indication for the start of moult during regression of ovary size, but the data were less robust (Figure 7.10).

In addition to the hormonal regulation of reproductive and moult functions, it is important to consider that energy and nutrients are needed to develop new feathers (Figure 7.11). Masman (1986) estimated that the energy needed to produce 1 g of feather is about 130 kJ and that a 30% increase in basal metabolic rate occurred during the most intense phase of moult. The energetic cost estimated by either indirect calorimetry or maintenance metabolism was higher during the moult period in captive

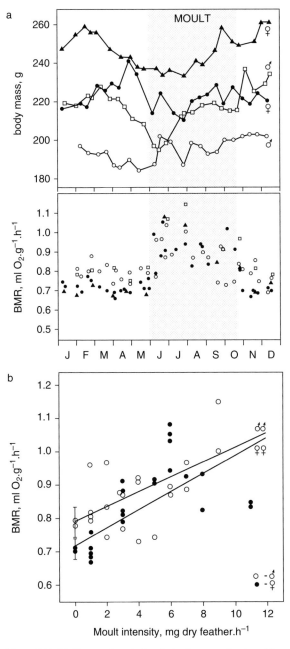

Figure 7.11 (a) Seasonal variation in body mass (top) and basal metabolic rate (bottom) in four individual kestrels. The grey bar indicates the average molting period for the four birds. (b) Mass-specific basal metabolic rate increases with the intensity of moult in both males (open circles) and females (solid circles). Reprinted with slight modifications from Dietz et al. (1992) with permission from the University of Chicago Press Journals.

kestrels (Dietz et al., 1992). This energetic cost also increased with the number of feathers to be produced (Dietz et al., 1992).

Several factors may contribute to the increased energetic demands faced while moulting. One factor that seems to be particularly important is the increased rate at which heat is being lost. Large heat loss would occur because of the reduced insulation as feathers are shed and for the increased number of blood vessels in the skin through which heat is lost (Village, 1990). That moult is costly and its timing has to be adjusted to other activities is further shown by the arrested moult, which occurs more frequently in males, probably because of the costs of hunting, which gives rise to trade-offs in the allocation of energy to different functions. Free-living kestrels seem to actually offset the energetic costs of moult by reducing their flight activity during the peak of moult (Masman et al., 1988). It might be that selection favoured the strategic reduction of flight activity during moult because this coincides with a period when prey are more abundant, so that kestrels may catch enough prey over a shorter period of time. While this may apply to northern populations, it is unclear if this explanation is plausible for populations living at lower latitudes, where voles are not the dominant prey as much as they are for northern populations.

7.9   Conclusions

Research on physiological and immune functions is fundamentally important to (i) understand the endogenous mechanisms that regulate given behaviours or the phenology of the life cycle and (ii) estimate costs underlying life-history trade-offs. The egg-laying period is the most energy-demanding activity of the reproductive phase for female kestrels. In contrast, the nestling-rearing period is the most demanding in terms of energy expenditure and generation of molecular oxidative damage for male kestrels. In nestlings, the metabolic and immune costs of sibling competition appear to be the major reasons for the trade-off between growth and self-maintenance.

Although energetic constraints might be important regulators of the timing of reproduction in kestrels inhabiting seasonal environments, they might be less important for kestrels breeding in less-seasonal environments, where the availability of prey (thus of sources of energy) is more constant across the year. Also, the capacity of kestrels to live under conditions of heat and aridity implies that they might rely on different strategies to tolerate or face elevated body temperatures and metabolic demands, which complicates further the generalisation of data collected from populations living at high latitudes. Physiological plasticity might also differ between migratory and year-round resident populations, implying that generalisation of findings is not straightforward.

We know very little on many other aspects of physiological and immune functions. For example, little research has been done on (i) the link between disease ecology and population dynamics; (ii) the many hormones (e.g. prolactin, melatonin, ghrelin, thyroid hormones) that regulate parental care, need of food, energy homeostasis or fledging time; (iii) mechanisms that regulate the cellular oxidative status and senescence, such as glutathione, antioxidant enzymes and telomere dynamics. Finally, it is important to recognise that ecological energetics relied heavily on metabolic rate and energy expenditure, which are very indirect metrics of energy production and use; this can significantly weaken our ability to link metabolic rate to organism performance or fitness.

8 Environmental Toxicology

8.1 Chapter Summary

Many types of chemical pollutants biomagnify across the food chain and reach their highest levels in predators like kestrels. In urban and suburban environments, kestrels are also being exposed to non-chemical pollutants (e.g. electromagnetic fields, light and noise pollution), which are becoming a growing concern. This chapter summarises the ways through which a range of chemical and non-chemical pollutants may influence the behaviour, physiology and reproduction of kestrels, and describes how patterns of population recovery have followed the control and withdrawal of some chemical pollutants.

8.2 Introduction

Rachel Carson's influential book *Silent Spring* (Carson, 1962) and reports on toxic chemicals by Stanley Cramp (Newton, 2013), which exposed in the 1960s the hazards of the widespread use of pesticides for wildlife, have greatly contributed to heighten public awareness of pollution problems. Misuse of pesticides and many other environmental pollutants is now widely recognised to threaten not only wildlife but human communities as well. In the last 15 years, the European Union has developed a range of legislative instruments and policies to limit and control environmental contamination in order to protect human and wildlife health. Moreover, several European countries have launched a number of environmental contamination biomonitoring programmes using raptors as sentinel species, such as the Predatory Bird Monitoring Scheme in the United Kingdom (Walker et al., 2008; Gómez-Ramírez et al., 2014).

Raptors are actually particularly prone to accumulate many types of chemical pollutants because of biomagnification, i.e. the increasing concentration of a substance in tissues of organisms at successively higher levels in a food chain. Birds of prey were actually among the main victims of pesticide usage in the 1950s. DDT and its breakdown products together with other chlorinated hydrocarbon pesticides posed a serious threat to conservation of several raptor species, including kestrels. Birds of prey may accumulate a little of these chemical pollutants per day without suffering any sudden toxic effects. Once accumulated in the body, most of the DDT is rapidly converted to its much more stable metabolite, DDE. Rather than killing the birds immediately after ingestion, at sublethal level, DDE reduces the availability of calcium carbonate during eggshell formation (e.g. Ratcliffe, 1967, 1970; Hickey & Anderson, 1968; Wiemeyer & Porter, 1970). As a consequence, the eggs were so fragile that eggshells simply broke up before hatching. The low reproductive success caused by eggshell thinning rapidly turned into population crashes of many birds of prey until the discovery that these chemicals were responsible for such a catastrophe. In the 1970s and 1980s, agricultural use of DDT was banned in most countries and bird populations started to recover quickly thereafter.

Kestrels are still exposed to many categories of chemical pollutants, which differ in propensity to bioaccumulate and biomagnify as well as in their mode of action and the biological effects they may cause. Chemical pollutants may cause a number of detrimental effects on the organism (e.g. reduction in reproductive success, increased mortality) or even alterations of the phenotype (e.g. body colourations; Figure 8.1) with unclear consequences for the fitness. Birds of prey may also be exposed to non-chemical pollutants, such as the low-frequency electromagnetic fields generated by power lines or noise and light pollution. Kestrels have played a central role as model taxa in experimental investigations of the behavioural and physiological consequences of pollution in free-ranging vertebrates. In the following paragraphs, we have illustrated a number of examples of how different categories of pollutants affect the behaviour, physiology and reproduction of kestrels.

8.3   Pesticides

Pesticides are substances that are made to control pest species (e.g. herbicides, insecticides, rodenticides, fungicides). Like other raptor species, kestrels are sensitive to the effects of pesticides. In the late 1950s, thousands of birds were regularly found dead in agricultural fields across Britain, where organo-chlorine pesticides (aldrin, dieldrin and heptachlor) were used to dress grains. Chemical analyses revealed the presence of high concentrations of organochlorine residues in their body tissues (Newton, 2013). The initial decline of populations of common kestrel in eastern England during the early 1960s coincided with this first widespread use of organochlorine pesticides as seed dressings (Village, 1990). Restrictions in pesticide usage started in 1962 and recovery of kestrel numbers began in 1965 (Cooke et al., 1982). Similarly, in Sweden, survival rates of adults based on recoveries of ringed birds were higher before than after the start of pesticide and alkyl-mercury use (Wallin et al., 1983; see also Wallin, 1984).

Although the use of many pesticides was banned long time ago, some are still present in the environment and can accumulate in kestrels (legacy pesticides). Female kestrels can detoxify themselves during the laying phase by depositing the chemicals accumulated in their body in their eggs. Therefore, eggs represent a good matrix to reveal local exposure in non-migrating birds and to allow comparison of pesticide

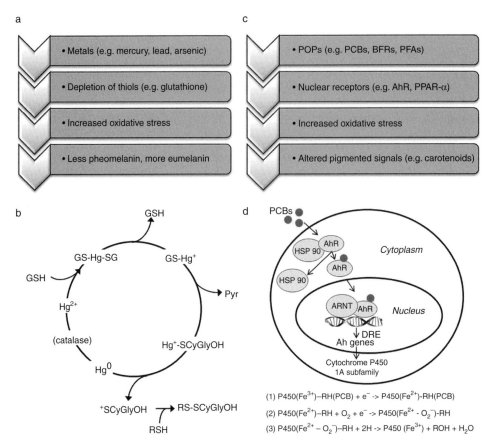

Figure 8.1 (a) Simplistic diagram of the hypothetical links among heavy metals, oxidative stress and melanin-based colourations via the glutathione pathway. (b) Schema of the catalytic cycle for the depletion of the cellular glutathione pool by mercury, proposed on the basis of mass spectrometric behaviour and biochemical evidence (Rubino et al., 2006); GSH, glutathione; GS-Hg+, mono-S-glutathionyl-mercury(II); Hg, mercury; CSCyGlyOH, oxidised cysteinyl-glycine; Hg+-SCyGlyOH, mercury(II)–cysteinyl-glycine conjugate; Pyr, pyroglutamic acid; RSH, thiols. (c) Simplistic diagram of the hypothetical links among persistent organic pollutants (POPs), oxidative stress and colouration. Different classes of POPs can generate oxidative stress via binding with specific cellular receptors, including the aryl hydrocarbon receptor (AhR) and proliferator-activated receptor-alpha (PPAR-α). (d) Schema of effects of dioxin-like pollutants (e.g. co-planar PCBs, PBDEs) via induction of the cytochrome P450 pathway (Regoli & Giuliani, 2014), which induces uncoupling of the catalytic cycle of the enzyme CYP1A allowing the heme iron within the active site of this enzyme complex to undergo cycles of oxidation and reduction, and act as a Fenton catalyst generating free radicals. HSP 90, heat shock protein 90; ARNT, aryl hydrocarbon receptor nuclear translocator; DRE, dioxin-responsive element. Different classes of pollutants can act on the same cellular targets altering the cell oxidative machinery. See Regoli and Giuliani (2014) for a full review on these aspects. Reprinted from Marasco and Costantini (2016).

levels among different areas. An analysis of 27 unhatched eggs collected from kestrel nest boxes in central Italy revealed striking differences between birds breeding in the countryside and those breeding along an urban portion of a polluted river in the city of Rome (Dell'Omo et al., 2008). Eggs collected in 1999 from two nests located near the banks of the Aniene river showed the highest concentrations of hexachlorobenzene (45.2 and 84.5 ng g^{-1} wet weight), aldrin/dieldrin (32.1 and 52.1 ng g^{-1} wet weight) and DDE (1306 and 2313 ng g^{-1} wet weight). In 2005, an entire clutch of four unhatched eggs was collected from one of those nest boxes near the Aniene river. The types of pesticide residues in eggs were qualitatively similar to those measured in eggs collected from the same nest box six years before; their concentrations were also high compared to those in eggs collected from nests in the countryside (Dell'Omo et al., 2008). The concentration of aldrin/dieldrin (from 44.5 to 64.5 ng g^{-1} wet weight) was comparable to that recorded in 1999; however, the concentrations of both hexa-chlorobenzene (from non-detectable to 7.6 ng g^{-1} wet weight) and DDE (from 1011 to 1416 ng g^{-1} wet weight) were lower than those recorded in 1999. Interestingly, we also observed nestlings with skeletal deformities in nest boxes near the Aniene river, suggesting possible negative effects of pesticides on the development of nestlings.

In addition to legacy pesticides, kestrels also accumulate relevant amounts of current-use pesticides. Rodenticide use is a common practice to minimise vole damage on croplands during vole outbreaks. However, rodenticides may cause problems to those raptor species that prey on voles (i.e. secondary poisoning). Martínez-Padilla et al. (2017b) confirmed that the use of the rodenticide bromadiolone in agroecosystems of Spain led to the secondary exposure of common kestrels (average concentration of 0.248 ng ml^{-1}) in croplands. They also found correlative evidence for a potential detrimental effect of bromadiolone on the body condition of nestling kestrels. Body mass was statistically and significantly lower in nestlings with bromadiolone residues than in nestlings without bromadiolone residues despite voles being abundant across the study areas (Figure 8.2). Specifically, nestlings with detectable levels of bromadiolone in the blood weighed 207 g on average, and those that did not weighed 222 g on average (Martínez-Padilla et al., 2017b).

8.4   Essential and Non-essential Elements

Accumulation in body tissues of large amounts of either essential (e.g. copper, iron, selenium, zinc) or non-essential (e.g. cadmium, lead, mercury, silver) trace elements may have dramatic consequences for individual health and fitness. Mercury contamination is a priority topic for the international community. The Minamata Convention on mercury, which aims to 'protect human health and the environment from anthropogenic emissions and releases of mercury and mercury compounds' (Article 1), was made in 2013 and entered into force in 2017 (16 August 2017, www.mercuryconvention.org/). This convention has been signed by 128 countries so far. Mercury biomagnification along food webs, mainly due to cumulative transfers of the methylated form, can lead to extremely high mercury concentrations in piscivorous species at the top of trophic networks, but also in terrestrial predatory birds (Eagles-Smith et al., 2018; Whitney & Cristol, 2018). Recent studies on both captive and free-

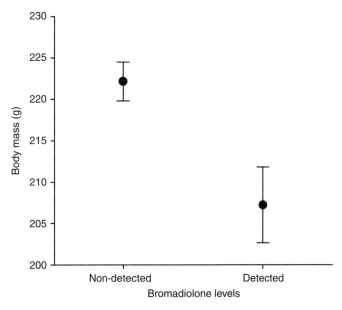

Figure 8.2 Body mass (mean ± standard error) of nestling common kestrels in relation to prevalence of the pesticide bromadiolone. Reprinted from Martínez-Padilla et al. (2017b) with permission from John Wiley and Sons.

ranging birds indicated that mercury may have a range of detrimental effects on neurobehavioural, endocrine, immune or reproductive functions (Whitney & Cristol, 2018). In American kestrels, experimental administration of environmentally relevant amounts of methylmercury reduced reproductive success (Albers et al., 2007), suppressed the cell-mediated immune function and caused inflammation (Fallacara et al., 2011), but a dose-dependent response was not apparent. For example, the detrimental effects of methylmercury on immune function was similar between groups having concentrations of total mercury in whole blood ranging from 3.82 to 8.72 $\mu g\ g^{-1}$ wet weight and from 36.9 to 62.0 $\mu g\ g^{-1}$ wet weight, respectively (Fallacara et al., 2011). In Sweden, decrease and recovery patterns of kestrel populations coincided with the use and ban of alkyl-mercury, respectively (Wallin, 1984).

8.5   Polychlorinated Biphenyls

Polychlorinated biphenyls (PCBs) are persistent and lipophilic environmental contaminants with the formula $C_{12}H_{10-x}Cl_x$. PCBs were once widely deployed as dielectric and coolant fluids in electrical apparatus, carbonless copy paper and in heat transfer fluids. PCBs were also used in many other applications, such as in inks, adhesives, rubber products, paints, pesticide fillers and plasticisers. Their production was banned by US federal law in 1978, and by the Stockholm Convention on Persistent Organic Pollutants in 2001.

Birds of prey may accumulate a large range of PCB congeners and suffer from their interference with endocrine mechanisms. Accumulation of PCBs in eggs may have detrimental consequences for the embryo. Hoffman et al. (1998) tested the effects of several doses of PCB 77 (0, 100 and 1000 ng g^{-1}) and PCB 126 (0, 0.23, 2.3, 23 and 233 ng g^{-1}) on embryos of American kestrel by air cell injections of each PCB, separately, on day 4 post-laying. Both doses administered of PCB 77 resulted in 23–31% mortality during the first 2 weeks following treatment and in 62–65% mortality by the time of pipping, in contrast to 27% for controls. Hatching successes for both treatment groups were 38% and 27%, which were significantly lower than the 65% hatching success of controls. The highest dose administered of PCB 126 resulted in only 17% mortality within the first 2 weeks following treatment and 71% mortality by the time of pipping, in contrast to 20% for controls and 31% for the 23 ng g^{-1} group. Malformations in embryos and hatchlings appeared to occur in a dose-dependent manner for both PCB congeners; the most frequent malformations included external yolk sac in hatchlings and shorter lower beak. Finally, the estimated median lethal doses (LD$_{50}$, dose required to kill half the members of a population after a given test duration) of PCB 77 and PCB 126 were 316 and 65 ng g^{-1} (1070 and 197 nmol kg^{-1}), respectively (Hoffman et al., 1998).

Exposure to PCBs may also be detrimental after hatching or even at adulthood. Captive American kestrels exposed to a daily dose of about 7 mg PCBs (dietary mixture of a 1 : 1 : 1 Aroclor 1254, 1248 and 1260) kg^{-1} body mass laid smaller clutches later in the season and laid more totally infertile clutches (Fernie et al., 2001). Hatching success was reduced in PCB-exposed pairs, and 50% of PCB nestlings died within 3 days of hatching. Nearly 60% of PCB-exposed pairs with hatchlings failed to produce fledglings (Fernie et al., 2001).

PCB-exposed kestrels also showed significant changes in their carotenoid-based colourations (Bortolotti et al., 2003), whose intensity is under endocrine regulation to some degree (Figure 8.3). At pairing, control males were significantly brighter than PCB males and in the short period between pairing and courtship, male colour became brighter for PCB birds, whereas it did not change significantly in control males (Bortolotti et al., 2003). In winter, PCB exposure resulted in patterns of colour variation opposite to controls: exposed adult males were duller and juveniles of both sexes were brighter than controls. Sexual dimorphism in colour was apparent in control adults, but not for PCB-exposed birds (Bortolotti et al., 2003).

PCB exposure could also diminish production of the thyroid hormones thyroxine (T4) and triiodothyronine (T3) or of corticosterone in American kestrels (Smits et al., 2002; Love et al., 2003b). In similar studies on American kestrels, Hoffman et al. (1996) found that the PCB 126 caused degenerative lesions of the thyroid, lymphocyte depletion of the spleen and bursa of Fabricius, reduced immune organ weight and caused higher susceptibility to oxidative stress in nestlings. Quinn et al. (2002) found that Aroclor 1242 suppressed levels of T4 in adults.

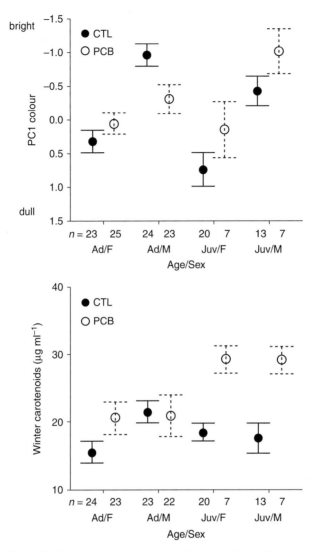

Figure 8.3 Colour values (mean ± standard error) by digital photographs (top) and plasma carotenoid concentration (mean ± standard error, bottom) as determined for different age/sex classes of PCB-exposed and control (CTL) American kestrels in winter. Reprinted from Bortolotti et al. (2003) with permission from John Wiley and Sons.

8.6   Polybrominated Diphenyl Ethers

Polybrominated diphenyl ethers (PBDEs) have been used as flame retardants in polymers, textiles, electronics and other materials because they have an inhibitory effect on combustion chemistry and tend to reduce the flammability of products containing them. These compounds are persistent, lipophilic, readily bioaccumulate in aquatic and terrestrial organisms and biomagnify in food chains. PBDEs may cause a range of potentially detrimental effects on the organism.

PBDEs may alter the immune function and structure of birds. Fernie et al. (2005) injected eggs of captive American kestrels with environmentally relevant concentrations of penta-BDE congeners -47, -99, -100, and -153 (18.7 mg per egg) and supplemented nestlings with the same PBDE mixture (15.6 ng g^{-1} body mass per day) until they were 29 days old. The body burden of PBDEs in treated birds was 86.1 ng g^{-1} wet weight, while that in control birds was just 0.73 ng g^{-1} wet weight. PBDE-exposed birds had a greater T-cell-mediated immunity stimulated by PHA injection, which was negatively associated with increasing BDE-47 concentrations, but a reduced antibody-mediated response that was positively associated with increasing BDE-183 concentrations (Fernie et al., 2005). There were also structural changes in the spleen (fewer germinal centres), bursa of Fabricius (reduced apoptosis) and thymus (increased macrophages), and negative associations between the spleen somatic index and total PBDEs, and between the bursa somatic index and BDE-47 (Fernie et al., 2005).

The 1,2-dibromo-4-(1,2-dibromoethyl)cyclohexane (DBE-DBCH, formerly TBECH) is an emerging flame retardant with androgen-potentiating ability and other endocrine disrupting effects in birds and fish. Marteinson et al. (2017) exposed captive American kestrels to environmentally relevant levels of β-DBE-DBCH (0.239 ng g^{-1} body mass day^{-1}), from 4 weeks before pairing until the nestlings hatched (mean 82 days) and compared them with control kestrels. β-DBE-DBCH was not detected in the tissues or eggs of these birds, nor were any potential metabolites found, despite the low limits of detection of the methods used, suggesting that β-DBE-DBCH may be rapidly metabolised and/or eliminated by kestrels (Marteinson et al., 2017). Exposed males had lower concentrations of total thyroxine (TT4) and higher concentrations of free thyroxine (FT4) and testosterone than control males; in contrast, exposed and control females had similar concentrations of TT4, FT4 and 17β-estradiol. These results indicate that exposure to a low concentration of β-DBE-DBCH may elicit effects on multiple hormones in the thyroid and steroid pathways, raising concerns about its impact on wildlife (Marteinson et al., 2017).

Alterations of thyroid hormones may have been due to the thyroid-disrupting action of PBDEs. Fernie and Marteinson (2016) showed that embryonic exposure to the PBDE mixture DE-71 altered the thyroid gland function because the production of both TT3 and TT4 following the intramuscular injection of the thyroid-stimulating hormone differed among treatment groups of female, but not of male, nestling American kestrels. However, the response of the thyroid gland was not dose-dependent, indicating complex and multiple direct and indirect ways through which the thyroid axis is affected by PBDEs (Figure 8.4).

Exposure early in development to flame retardants may also cause alterations in brain development. *In ovo* injection at embryonic day 5 of several doses of either bis (2-ethylhexyl)-2,3,4,5-tetrabromophthalate (BEH-TEBP) (13, 64, or 116 µg g^{-1} egg) and 2-ethylhexyl-2,3,4,5-tetrabromobenzoate (EH-TBB) (12, 60, or 149 µg g^{-1} egg), which are components of FireMaster 550® and 600® (a mixture of flame retardants),

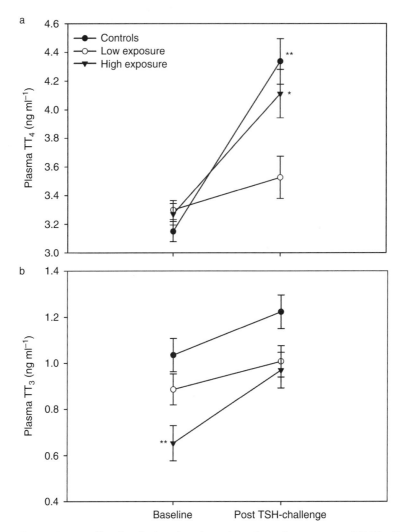

Figure 8.4 In nestling female American kestrels, embryonic exposure to DE-71 affected: (a) circulating total thyroxine (TT4) concentrations overall and the temporal increases in plasma TT4 concentrations following thyroid-stimulating hormone (TSH) challenge; and (b) circulating total triiodothyronine (TT3) concentrations overall and modestly influenced the increase in TT3 levels following thyroid-stimulating hormone (TSH) challenge. Baseline TT3 concentrations were lower in high-exposure female nestlings than control females. *$p < 0.05$; **$p < 0.01$. Reprinted with slight modifications from Fernie and Marteinson (2016) with permission from John Wiley and Sons.

caused sex-specific effects on brain structure at hatching in American kestrels (Guigueno et al., 2018). The hippocampus was significantly enlarged in high-dose females compared to control females but smaller in low-dose females than in the other females. There was no significant effect of EH-TBB on hippocampus volume in hatchling female kestrels or of either chemical in male hatchlings and no effects of these concentrations of EH-TBB or BEH-TEBP on telencephalon volume or the level of symmetry between the hemispheres of the brain (Guigueno et al., 2018).

8.7   Perfluoroalkyl and Polyfluoroalkyl Substances

Perfluoroalkyl and polyfluoroalkyl substances (PFASs) are synthetically manufactured chemicals, produced since the 1950s, that are widely used for numerous industrial and commercial purposes as water repellents and surfactants (e.g. impregnation agents for carpets, papers and textiles, fire-fighting foam, non-stick coating and waterproof clothing; e.g. Jensen & Leffers, 2008; Muir & De Wit, 2010). PFASs are proteophilic and are highly persistent in the environment, bioaccumulate in living organisms and biomagnify along the food webs (e.g. Jensen & Leffers, 2008; Muir & De Wit, 2010). PFASs have raised many concerns about their potential physiological disrupting properties and negative impacts on reproductive fitness in wildlife, particularly in seabirds (e.g. lesser black-backed gull *Larus fuscus* in Bustnes et al., 2008; black-legged kittiwake *Rissa tridactyla* in Blévin et al., 2017 and Costantini et al., 2019). Birds of prey, including kestrels, may also accumulate significant amounts of PFASs. Perfluorooctane sulfonic acid (PFOS) was the predominating toxicant detected in eggs of Swedish common kestrels, accounting for 45% of total PFASs (Eriksson et al., 2016). Linear PFOS was the dominant PFOS (88%); it was highly enriched compared to commercial mixtures, suggesting a potential preferential accumulation of linear isomers (Eriksson et al., 2016). Perfluoroundecanoic acid (PFUnDA) was the predominating (16% of total PFASs) perfluorinated carboxylic acid (Eriksson et al., 2016). Although these results indicate that PFASs might be a concern, there are not yet any experimental investigations on the effects of ecologically relevant doses of PFASs on the behaviour and physiology of kestrels.

8.8   Electromagnetic Fields

Kestrels often breed on abandoned corvid nests or in nest boxes located on the pylons of overhead power lines (Chapter 4). In so doing, both adult and nestling kestrels are exposed to low-frequency (50–60 Hz) electric and magnetic fields (EMFs) generated by power lines. Fernie et al. (2000) reported that laboratory exposure of captive American kestrels to 60-Hz EMFs (30 mT, 10 kV m^{-1}) reduced eggshell thickness and hatching success, but increased fertility, egg size and fledging success. EMF-exposed nestling kestrels were heavier and had longer tarsi than unexposed nestling kestrels, while the growth of the ninth primaries and central rectrices were unaffected by EMF exposure (Fernie & Bird, 2000). Short-term exposure to EMFs also caused stimulatory and suppressive effects on melatonin production (Fernie et al., 1999) and reduction of total proteins, haema-tocrits, erythrocytes, lymphocytes and carotenoids (Fernie & Bird, 2001). These studies exposed the birds to constant EMFs, but the intensity of EMFs varies

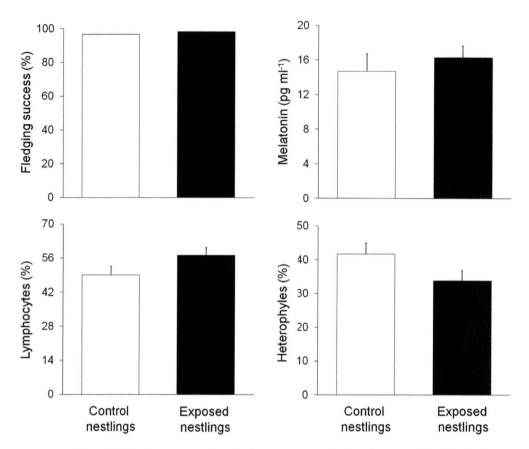

Figure 8.5 Nestling common kestrels that were exposed to low-frequency (50–60 Hz) electric and magnetic fields (EMFs) generated by power lines during the development period had similar probability to fledge, serum melatonin concentration, lymphocytes and heterophyles to those that were not exposed to EMFs. Data from Dell'Omo et al. (2009).

within and between days depending on the actual needs for electric current. Studies carried out on nestling common kestrels raised in nest boxes attached to utility line pylons did not find any evidence for short-term effects of EMF exposure on a range of traits. Costantini et al. (2007a) showed that the average values of EMF magnitude within nest boxes varied from 0.20 to 20.44 μT and that these values were not correlated to those of a serum marker of oxidative damage, a serum marker of non-enzymatic antioxidant capacity, the serum concentration of total carotenoids or the body condition index of nestlings. Moreover, the values of each metric recorded in EMF-exposed nestlings were similar to those recorded in unexposed nestlings. Another study also did not record any difference between EMF-exposed and -unexposed common kestrel nestlings in serum melatonin, immune cell counts, erythrocytes, thrombocytes, growth and survival until fledging (Figure 8.5; Dell'Omo et al., 2009).

8.9   Conclusions

Kestrels are sensitive to the toxic effects of a number of chemical pollutants. This has been demonstrated by many experimental investigations carried out on animals maintained under captivity conditions that were exposed to realistic concentrations of given pollutants. However, these results need replication under free-living conditions, where birds are often exposed to a cocktail of pollutants and to a number of environmental stressors that may amplify or reduce the toxic effects of each type of pollutant. It is also unclear whether any stimulatory effects caused by a low-dose exposure occur and have any positive effects on fitness as predicted by hormetic models (Mattson & Calabrese, 2010).

Kestrels are also exposed to non-chemical pollutants. As illustrated in the present book, kestrels may breed on pylons of electric companies, where they are exposed to variable magnitudes of electromagnetic fields. Kestrels have also became common birds in urban areas, where exposure to both light and noise pollution may be intense. We know very little about the fitness effects of exposure to non-chemical pollutants, thus we urge more research in this area.

9   Movement Ecology

9.1   Chapter Summary

Common kestrels are defined as partial migrants because they have variable migratory strategies over their geographic distribution, from obligate migrants in the north of Europe to more sedentary habits in central and southern regions. Migratory strategies are subject to a multiplicity of external and internal drivers, which are still not well understood. Many individual kestrels also disperse, rather than migrate, from the breeding or birth area. Dispersal distances are longer in females than in males and in yearlings than in older individuals. The dispersal is influenced by a number of factors, such as individual propensity and food availability. The deployment of GPS data-loggers and geolocators on kestrels will greatly improve our understanding of their movement ecology and help to discriminate between migration and dispersal.

9.2   Introduction

Most knowledge on the movement ecology of common kestrels has been gathered by studies in central and northern Europe. This work is based on the analysis of recovery of birds ringed under various national ringing schemes. In contrast, fewer long-term ringing studies have been carried out in southern European countries, meaning that much less is known about the movement ecology of kestrels at lower latitudes or on other continents (Figure 9.1).

Data collected from the recovery of ringed birds suffer some unavoidable biases that may hinder interpretation. For example, birds are more likely to be recovered in densely populated areas or in those areas where hunting activity is stronger. In recent years, the development of sophisticated modern techniques based on GPS tracking (e.g. GPS tags, which record the animal's location on Earth) has greatly improved our understanding of movement ecology of many animal species.

The possibility of deploying GPS instruments on kestrels has been constrained until recently by the small size of the birds. Small archival tags became available in the last decade, but the need to recapture the birds has limited the use of these tags to lesser kestrels, which are easier to capture at the nest than common kestrels. Also, interest in

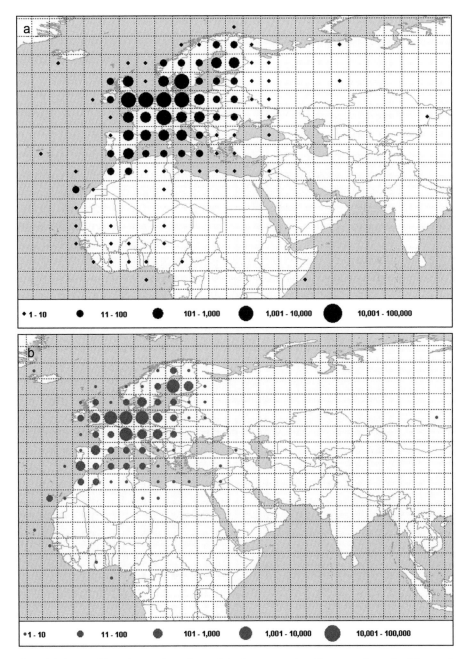

Figure 9.1 Maps of locations and numbers of ringing recoveries of common kestrels: (a) records of dead recoveries; (b) records on live recaptures/sightings. Data from du Feu et al. (2009).

tracking lesser kestrels has always been greater than common kestrels because lesser kestrels are truly migratory birds. Lesser kestrels have actually been tracked during migration to Africa with geolocators (Catry et al., 2011) and satellite tags (Liminana et al., 2012). In Chapter 3, we showed that small GPS-loggers can also be successfully deployed on common kestrels, opening new avenues for more detailed research on the movement ecology of this species.

9.3   Migration

The movements of kestrels out of the breeding area, as in many other birds, are generally driven by the need to find more favourable trophic conditions. Depending on the spatial scale and the uni- or bi-directionality of the movement, we refer to migration, when birds move seasonally along given directions (in general north–south and vice versa) twice a year over long distances, or to dispersal, when individuals move along various directions over shorter distances between the place of birth and the breeding site or between breeding sites. These two types of movements are also distinct at the temporal scale because they differ in the timing at which they occur and the number of individuals involved. Migration is seasonal and synchronous, and involves a large number of individuals, whereas dispersal involves individuals that spread randomly and singly with no temporal coordination with others (partners, parents and siblings), which are left behind in the breeding area (Figure 9.2 and Figure 9.3).

In northern Europe, kestrels are almost entirely obligate migrants, while they show increasingly sedentary habits (*facultative migrants*) in southern regions (e.g. Snow, 1968; Village, 1990; Adriaensen et al., 1998; Bauer et al., 2005). Village (1990) suggested that birds above the limit of permanent snow cover must migrate to the south in order to survive the winter. He sketched the line across Poland and Ukraine to separate the truly migrating populations of the north from those at the south of this line, which can be considered as partial migrants. At that time, however, satellite imagery was not developed as it is nowadays. The progression of the snow cover over winter can vary from one year to the next and in some years it can reach more southern limits. Therefore, by moving away from snow, kestrels can vary the extension of their winter movements from one year to another. Some European kestrel populations may show a leap-frog migration (Wallin et al., 1987; Village, 1990), i.e. autumn movement by the northern breeding birds to winter quarters, which lie further to the south than those occupied by the southern breeding birds. There is a considerable spread in the timing of migration within populations, as well as in the propensity to migrate or to stay around the breeding ground all year round. Also, migration strategies differ between either males and females or adults and young.

In central and southern Europe, it is difficult to distinguish whether kestrels, particularly the young, are on migration or dispersal as the local birds may move and can be replaced by birds coming from the north. This scenario becomes more complex by the fact that movements of many kestrels from central and southern Europe are too short to be considered as migration or too long to be considered as dispersal. Re-encounter data from ringed birds showed that kestrels can cover distances of thousands of kilometres. For example, individuals ringed (i) in Switzerland were re-encountered in Liberia and

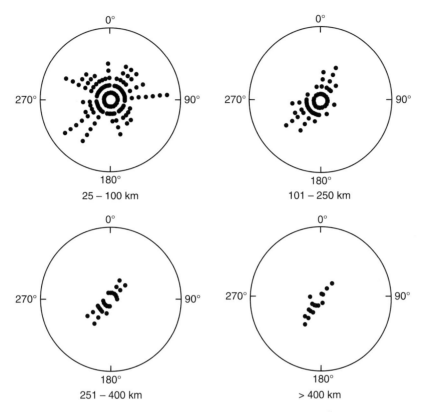

Figure 9.2 Direction of movement of Belgian kestrel nestlings, recovered before the end of March of the following year. Data are grouped in 10° classes, every dot represents one individual (Adriaensen et al., 1997). Reprinted with permission from the British Trust for Ornithology.

Madeira (Schifferli, 1965), (ii) Germany and Czech Republic were re-encountered in Nigeria and Ghana, respectively (Moreau, 1972), (iii) Germany and Belgium were re-encountered in Senegal (Adriaensen et al., 1997; Holte et al., 2016) and (iv) Germany and Sweden were re-encountered in Mali and Sierra Leone (Mebs & Schmidt, 2006).

Holte et al. (2016) analysed ringing and re-encounter data of 3070 common kestrels marked as nestlings between 1924 and 2011 in Germany (Figure 9.4). The longest distances travelled by a juvenile between ringing and re-encounter sites varied from 2367 to 4548 km. Overall, juveniles and females covered longer distances than adults and males, respectively. Migration was initiated mostly in September/October by juveniles, while in August movements seemed to mostly reflect dispersal. Thus, as previously suggested for kestrels inhabiting other European countries (e.g. Belgium in Adriaensen et al., 1997; Czech Republic in Riegert & Fuchs, 2011), German kestrels may also be considered as partially migratory and differential migrants (i.e. showing different migratory strategies), with some birds remaining in continental Europe and some individuals truly showing a migratory behaviour by wintering in the Mediterranean region.

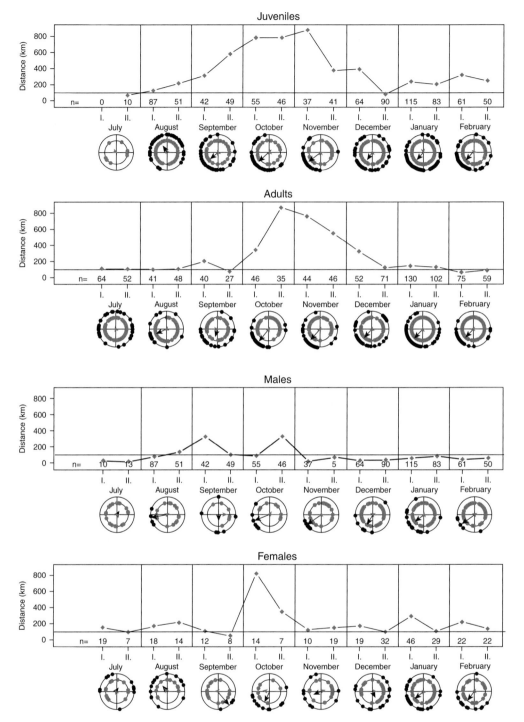

Figure 9.3 Distances and directions of re-encounters to ringing site during autumn migration and wintering of juvenile, adult, male and female common kestrels. Distances: black points = third quartiles of distances, *n* = number of all re-encounters per half month, grey line = 100 km. Directions: inner circles < 100 km, outer circles ≥ 100 km. Arrows represent mean directions with resultant vector lengths. Reprinted from Holte et al. (2016) with permission from the Museum and Institute of Zoology, Polish Academy of Sciences.

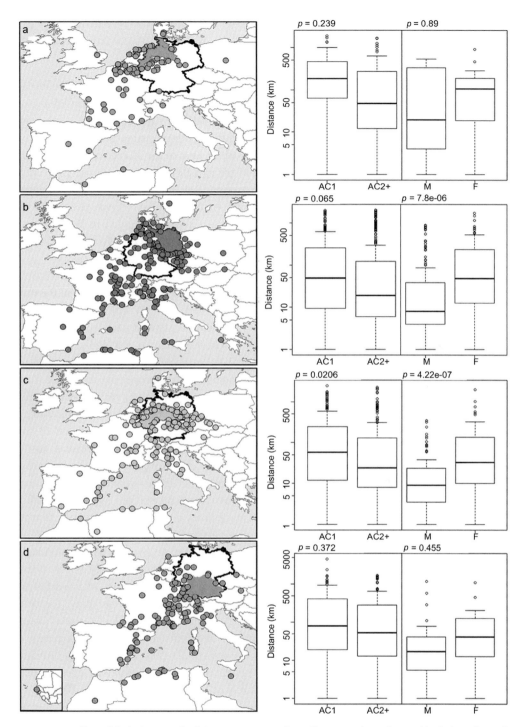

Figure 9.4 Autumn and winter re-encounter sites of common kestrels outside their regions of natal origin (left) and boxplots of re-encounter distances by sex and age classes (right). Reported *p* values result from comparisons of respective classes using negative binomial generalised linear models with no predictors. Note the log-scale on *y*-axes. A = northwest German lowlands, B = northeast German lowlands, C = central German highlands, D = southwest German highlands including Alps and the Foreland of the Alps. Boxplots: horizontal bar = median, whiskers = last observation within 1.5 times the interquartile range, circles = observations outside whiskers. AC1 = juveniles, AC2+ = adults. Reprinted from Holte et al. (2016) with permission from the Museum and Institute of Zoology, Polish Academy of Sciences.

The partial migration is a common feature of many populations across all countries in central Europe. For example, a study of recoveries of kestrels ringed in Poland or recorded in Poland but ringed abroad over the period 1931–2001 showed that most adult birds were recovered in the country during both migration periods (only three birds were recovered abroad; Śliwa et al., 2010). Five of the 32 recoveries of kestrels ringed in Poland as nestlings or as individuals in their first year of life were recorded outside the country. The majority of recovered birds that were ringed abroad came from Finland and other northern or north-eastern regions (Śliwa et al., 2010). Similarly, an analysis of about 1200 recoveries of kestrels ringed in Belgium and Luxemburg showed that the majority of nestlings dispersed in all directions and a low proportion of birds showed a truly migratory behaviour (Dhondt et al., 1997).

British kestrels are also partial migrants. A study of 213 recoveries of 1982 kestrels (a recovery rate 12.4%) ringed as nestlings up to 1956 showed that birds in the first autumn (when 70% of recoveries occurred) dispersed randomly within a short distance from the nest of origin (Thomson, 1958). A few birds showed a tendency to migrate towards the south beyond the borders of Great Britain: a few individuals were recovered in Belgium (3), France (13) and the north of Spain (1). In the north of Great Britain, kestrels appear to have a stronger migratory propensity than those from the south of Great Britain, and young birds generally tend to cover longer distances than adults (Snow, 1968). Dispersal of young kestrels in late summer was followed by a predominantly south-east migration, with a return to the neighbourhood of the birthplace in the following March–April. Those kestrels that migrated overseas mostly flew down through west-central France towards the western end of the Pyrenees. Movement to the very south is uncommon; a kestrel from Ayrshire (Scotland) was found in Tenerife and another from north-east Scotland reached Morocco (Riddle, 2007). The northward movement in spring follows a more easterly course. Ireland appears to be a subsidiary wintering area for northern British kestrels.

In Spain, kestrels are also considered as partial migrants because some individuals migrate in autumn, while others stay locally all year round. Interestingly, both resident and non-resident kestrels changed their feeding habits during autumn, but in a different way. Resident kestrels substituted grasshoppers, a typical summer prey, for field crickets and/or mammals (typical winter prey), whereas non-resident kestrels hunted fewer crickets and mammals, but more mantids and flying ants, which are not available later, during winter (Aparicio, 2000). Also, non-resident kestrels had a larger trophic diversity but a lower biomass per whole pellet than resident kestrels, suggesting that birds unfamiliar with the territory might rely on different prey (Aparicio, 2000).

It is evident that kestrels also show significant individual variation in migratory tactics. Village (1985a) observed that during the spring migration males and females arrived over roughly the same period; however, within pairs, the males arrived about 4 days, on average, before their females. This variation in migratory phenology had consequences for reproductive activity. Migrants that arrived early tended to pair and lay earlier than those that arrived late; later-pairing birds partly compensated for this by having a shorter courtship period than those that paired earlier (Village, 1985a).

9.4   Dispersal

Dispersal is classically distinguished in two different categories: *natal dispersal* refers to the movement of juveniles from the natal site to the site of first breeding; *breeding dispersal* refers to the movement of individual birds that reproduced previously between subsequent sites of reproduction (Greenwood & Harvey, 1982). Both natal and breeding dispersal are influenced by the distribution of suitable habitat and nest sites; it can influence survival, mating success, and reproductive rates of individuals, thus linking their behaviour to population dynamics.

9.4.1   Natal dispersal

The process of natal dispersal includes the so-called *post-fledging dependence period*. This is defined as the period between the first flight outside the nest and independence from the parents (Mock & Parker, 1997). It is therefore a critical life-history phase during which the young must learn and develop the foraging skills to become self-sufficient (Bustamante, 1994; Bustamante & Negro 1994). There is little information on the post-fledging dependence period of the common kestrel across its range. The duration of the post-fledging dependence period appears to depend, on one hand, on the ability of young birds to acquire hunting skills and, on the other hand, on the behaviour of the parents (e.g. amount of food provisioning, parental aggression; e.g. Tinbergen, 1940; Bustamante, 1994; Vergara et al., 2010; Boileau & Bretagnolle, 2014). This suggests that the post-fledging dependent period is determined by a parent–offspring conflict because, from the offspring's perspective, selection would favour extension of the post-fledging dependent period if this increases their survival probability; however, from the parental perspective, an extended post-fledging dependent period might come at a cost in terms of reduced future reproductive outcomes (e.g. López-Idiáquez et al., 2018). Although not always observed, competition among siblings might also contribute to determine the duration of the post-fledging dependent period (Vergara & Fargallo, 2008b). Field observations showed that, after fledging, the young remain near the nest for some time during which they make longer and longer flights and gradually acquire several skills. Parents stimulate their offspring, providing food to those that first reach them on a perch, on the ground or in mid-air (Village, 1990). Parents also progressively decrease the provisioning of food in order to stimulate offspring to hunt on their own. The young need time to acquire hunting skills, thus initially their diet is rich in insects, which are much easier to catch than small vertebrates.

Ringing recoveries of common kestrels from north-western Europe showed that a very low proportion of birds is migratory, and that even large-distance movements can be attributed mainly to dispersal movements of young birds (Adriaensen et al., 1998). In Great Britain, young kestrels disperse from July onwards fairly quickly (Village, 1990). In the first phase of dispersion, more than 70% of recoveries were within 75 km of the ringing site (Village, 1990). Also, birds born earlier in the season tended to disperse less than those born later; early-born young were five times more likely to breed locally when yearlings than were late-born young in both Great Britain and the Netherlands (Dijkstra, 1988; Village, 1990). Also, young birds tended to disperse farther in years of lower food availability (e.g. Adriaensen et al., 1998).

In south-western Spain, fledglings remained at the breeding site an average of 16 days and started to learn how to hover at around 10 days after fledging (Bustamante, 1994). There was large individual variation in the post-fledging dependence period (47 days), possibly due to variable efforts in food provisioning of parents. Vergara et al. (2010) found that food supplementation increased the length of the post-fledging dependence period: young from food-supplied nests became independent later. Similarly, Boileau and Bretagnolle (2014) concluded that the post-fledging dependence period might mainly be under parental control in kestrels. In a coastal area in western France, they found that the duration of post-fledging dependence period lasted on average 18 days with a range of 3–31 days. Fledglings in better body condition stayed near the nest longer. The average daily distance from the nest increased about 15 m per day until independence. At the end of the post-fledging dependence period, the mean distance from the nest was 372 m (range 50–1000 m); after independence, fledglings rapidly dispersed away from the area (Boileau & Bretagnolle, 2014).

In other kestrel species, the post-fledging dependence period is longer than that recorded in the common kestrel. In the rock kestrel (considered as a subspecies by some authors, Chapter 1), Komen and Myer (1989) found that the post-fledging dependence period was 41.5 days in an urban environment of Namibia. In the Madagascan banded kestrel, Rene de Roland et al. (2005a) observed that the young remained near the nest up to 40 days of age and were completely independent at around 56 days of age. The quantity of prey items delivered by the adults decreased after 41 days of age. Prey-capturing attempts of young started at 49 days of age and continued until the first young was observed catching its first small chameleon at 52 days of age, after which prey deliveries of parents decreased dramatically (Rene de Roland et al., 2005a).

9.4.2   Breeding Dispersal

The analysis of a Finnish database (period 1985–2008) showed that the maximum observed breeding dispersal distances of common kestrels were 152 km for males (ignoring one exceptional 715 km) and 194 km for females (Figure 9.5). Thus, females dispersed longer distances than males and yearlings dispersed twice as much as the distances covered by older individuals (mean values: yearling males, 16.9 km; older males, 7.9 km; yearling females, 55.0 km; older females, 22.4 km; Vasko et al., 2011).

Individuals showed some degree of consistency of dispersal distances across years; successive dispersal distances of the same individual were significantly correlated both in males ($r_s = 0.34$, $n = 108$, $p = 0.0004$) and females ($r_s = 0.39$, $n = 78$, $p = 0.0005$). Moreover, kestrels dispersing over a short distance at the first observed dispersal event

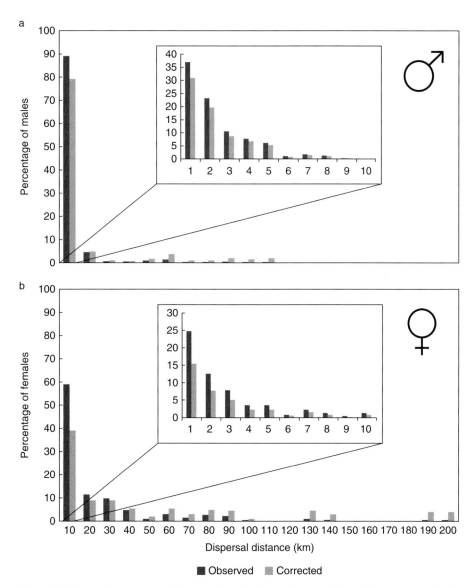

Figure 9.5 Observed and corrected dispersal distributions of male and female kestrels breeding in Finland during the period 1985–2008. Corrected numbers are calculated by dividing observed numbers in each distance class by trapping probability in that class. Reprinted from Vasko et al. (2011) with permission from John Wiley and Sons.

tended to have a higher probability of being re-trapped in the study area after their successive dispersal events (Vasko et al., 2011). These results suggested that the dispersal distance might be regulated to some extent by individual endogenous features that could lead to a distinction between short- and long-distance dispersers.

Dispersal distances also depended to some degree on vole abundance (Figure 9.6). The longest dispersal distances were most likely to occur in the decrease and low

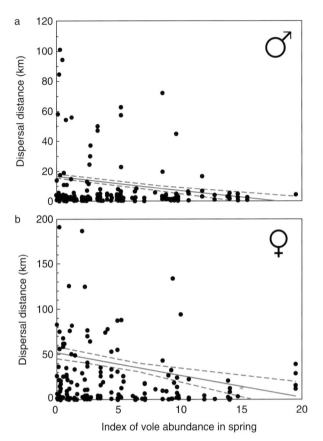

Figure 9.6 Dispersal distances of male and female kestrels in relation to vole abundance in the spring of settlement. Regression lines are shown with 95% confidence interval. Reprinted from Vasko et al. (2011) with permission from John Wiley and Sons.

phases of the vole cycle because the probability of finding a food patch probably increased with search distance (Vasko et al., 2011). However, the effect of vole abundance on dispersal was somewhat different between sexes. Females exhibited more than males a condition-dependent dispersal strategy, in which the individual dispersal distances were more strongly affected by the temporal fluctuation in vole abundance. Females in better body condition dispersed further than those in worse body condition, whereas there was no correlation between male body condition and dispersal distance (Terraube et al., 2015). These results suggested that the costs and benefits of staying closer to or dispersing further from the breeding site might significantly differ between males and females. Terraube et al. (2015) suggested that longer dispersal distances might have reproductive costs for the males but reproductive benefits for the females. For males, it might pay more in evolutionary fitness terms to stay closer to the breeding ground (i.e. philopatric behaviour) to avoid costs of acquiring a new territory and getting familiar with it whether conditions at the breeding ground are not too harsh that would drastically reduce the survival probability over the wintertime. This 'should I stay or should I go?' strategy might also be affected by the prior reproductive success; the probability of dispersing would increase after a poor reproductive season (Terraube et al., 2015).

9.5   Conclusions

Ringing programmes across European countries enabled demonstration that the movement ecology of common kestrels varies from north to south. In the very north of Europe kestrels are obligate migrants, whereas they are partial migrants in the central and southern regions, with migrants and sedentary individuals coexisting within the same populations. Many individuals disperse rather than migrate. This is particularly true for young kestrels, but adults also disperse at the end of the breeding season. The distinction between migration and dispersal is not always straightforward.

We know little about the causes and consequences of individual variation within populations in migratory or dispersal strategies and how they vary across years depending on environmental conditions and density-dependent mechanisms. For example, Holte et al. (2016) observed a significant decline of migratory movements of German kestrels after 1970. Although the authors did not exclude that these effects could be related to the intensification of agriculture in some regions, they also suggested that this decline could be related to climate change. This is because the number of resident birds was larger in years characterised by high values of the NAO index in summer, which leads to warmer and drier conditions in northern and western Europe, as well as in parts of central Europe. Thus, a 'decision' to migrate or not might also be made in response to the weather conditions experienced (Holte et al., 2016). Recent work hypothesised that individual differences in key personality traits affecting dispersal could have played a direct and active role in moulding the movement ecology and geographic distribution of many animal species (Canestrelli et al., 2016). We urge studies that investigate the role of personality in determining the individual propensity to disperse or migrate.

The study of movements of kestrels will be facilitated in the coming years by the availability of smaller and more sophisticated tracking devices. For example, the deployment of geolocators on lesser kestrels enabled detection of individual variation in the migratory movements of Portuguese birds, which wintered in Senegal, Mauritania or Mali (Catry et al., 2011). A subsequent study using satellite tracking confirmed that lesser kestrels from the same colony wintered in a wide area in western Sahel and revealed that birds covered longer daily distances during the spring migration because they were frequently flying at night (Liminana et al., 2012). There are also convenient new genetic approaches, such as the multilocus genotyping of moulted feathers, that has recently been used to determine the geographic origin of lesser kestrels along their migratory routes (Bounas et al., 2018).

10 Conservation Status and Population Dynamics

10.1 Chapter Summary

The common kestrel is evaluated as Least Concern at global level. However, at European level, the species is considered of conservation concern due to a continuous moderate decline since the 1980s due to agriculture intensification, landscape simplification, pesticide use, and loss of nesting sites. Moreover, the conservation status of some subspecies of common kestrel appears problematic. This chapter discusses the conservation status of kestrel species and subspecies, and the main top-down and bottom-up factors that affect the viability and stability of their populations. It also points out the strong limitations of our knowledge of the density-dependent and independent processes that regulate the demography and dynamics of kestrel populations. Important conservation-related topics, such as urbanisation, pesticides, or use of artificial nest boxes, have been discussed in detail in prior chapters.

10.2 Introduction

The conservation status of kestrel species is apparently good, with 11 species being evaluated as Least Concern, one as Vulnerable and one as Endangered (Table 10.1). However, the data are not exhaustive because surveys have covered only part of species' ranges and adequate estimates of population sizes have not always been produced. Although most species are evaluated as Least Concern, population declines have been recorded for several species, which calls for an urgent need to understand the causes of such declines in order to find solutions to slow down or stop them. Moreover, the conservation status of some subspecies raises some concern.

The main threats to the conservation status of the common kestrel include agricultural intensification and the widespread use of pesticides (Chapter 8). Although some studies showed that kestrels increased in frequency in more urbanised areas (Sorace & Gustin, 2010; Jokimäki et al., 2018), several recent studies have also shown that kestrels may not perform well in some urban environments compared to rural environments (Chapter 3). The use of artificial nest boxes in breeding programmes has been very important in increasing the availability of breeding sites in human-altered landscapes (Chapter 4). Captive breeding programmes coupled with reintroduction have also proved successful to restore species that were on the brink of extinction.

Table 10.1 IUCN conservation status, estimated population size of adult individuals and population trend. Data collected from BirdLife International (2016a–m).

Species	IUCN status	Estimated population size	Population trend
Common kestrel *Falco tinnunculus*	Least Concern	819,000–1,210,000 in Europe 4,000,000–6,500,000 in the world	Decreasing
Lesser kestrel *Falco naumanni*	Least Concern	61,000–76,100 in Europe	Stable
Fox kestrel *Falco alopex*	Least Concern	670–6700	Stable
Seychelles kestrel *Falco araeus*	Vulnerable	920–960	Stable
Grey kestrel *Falco ardosiaceus*	Least Concern	Unknown	Stable
Dickinson's kestrel *Falco dickinsoni*	Least Concern	Unknown	Stable
Moluccan or spotted kestrel *Falco moluccensis*	Least Concern	Unknown	Increasing
Madagascar kestrel *Falco newtoni*	Least Concern	Unknown	Increasing
Mauritius kestrel *Falco punctatus*	Endangered	170–200	Decreasing
Greater kestrel *Falco rupicoloides*	Least Concern	Unknown	Stable
Madagascar banded kestrel *Falco zoniventris*	Least Concern	670–6700	Stable
Australian nankeen kestrel *Falco cenchroides*	Least Concern	Unknown	Increasing
American kestrel *Falco sparverius*	Least Concern	Unknown	Stable*

* A recent report of McClure et al. (2017) found a significant decline of American kestrels across much of Canada and the USA.

10.3   IUCN Conservation Status

The most recent assessment of the conservation status of the common kestrel made by BirdLife International (2016l) has evaluated the species as Least Concern. This choice has been made because the species does not approach the thresholds to be classified as Vulnerable under the range size criterion (extent of occurrence < 20,000 km^2 combined with a declining or fluctuating range size, habitat extent/quality, or population size and a small number of locations or severe fragmentation) because of its very large geographic range. In Europe, there has been a significant moderate decline of 26% over the period 1980–2015, which prompted consideration of the species as European conservation concern SPEC3 (Figure 10.1 and Table 10.2; BirdLife International, 2015, 2017). In Burkina Faso, Mali, Niger and northern Cameroon the species decreased by 75–94% in all areas adequately surveyed between 1969–1973 and 2000–2004 (Thiollay, 2007). These declines have not been believed to be sufficiently rapid to approach the thresholds for Vulnerable under the population trend criterion (> 30% decline over 10 years or three generations). The population size is also

Table 10.2 National data on population sizes and trends. The quality of the population size and trend data is conveyed by one of three codes: 'italic' font denotes reliable quantitative data for the whole period and country; 'normal' font signifies generally well-known, but only poor, outdated or incomplete data available; '(bracketed)' indicates poorly known, with no quantitative data available. Countries are not reported when there are no records documenting presence of kestrels. Data collected by BirdLife International (2017).

Country	Number of individuals	Population trend since 2000
Albania	1000–3000	Decrease
Andorra	70–180	Unknown
Armenia	500–800	Unknown
Austria	15,000–24,000	Stable
Azerbaijan	(2000–10,000)	Unknown
Belarus	2400–3400	Stable
Belgium	8400–13,000	Stable
Bosnia and Herzegovina	(6000–8000)	Unknown
Bulgaria	8800–19,200	Stable
Croatia	(18,000–20,000)	Unknown
Cyprus	6000–10,000	Stable
Czech Republic	13,140–18,980	*Decrease*
Denmark	3000	*Decrease*
Estonia	1200–1800	Increase
Finland	*15,400–17,400*	*Increase*
France	*88,000*	Decrease
Georgia	3000–6000	Stable
Germany	*88,000–148,000*	*Fluctuating*
Gibraltar	*16–22*	*Stable*
Greece	16,000–24,000	Stable
Hungary	12,400–14,800	Unknown
Ireland	24,200–84,880	*Stable*
Italy	(16,000–24,000)	Increase
Kosovo	500–600	Unknown
Latvia	*256–506*	Unknown
Liechtenstein	30–50	Stable
Lithuania	600–800	Increase
Luxembourg	1000–1400	Stable
Macedonia	(3000–5000)	Unknown
Moldova	300–400	Decrease
Montenegro	800–1200	Unknown
Netherlands	6920–10,380	*Decrease*
Norway	(4000–8000)	Fluctuating
Poland	*9800–10,200*	*Stable*
Portugal	5000–15,000	Unknown
Romania	(40,000–100,000)	(Decrease)
Russia	(100,000–150,000)	Unknown
Serbia	8000–10,000	Stable
Slovakia	12,000–20,000	Stable
Slovenia	4000–5000	Unknown
Spain	*45,000–68,000*	*Decrease*
Sweden	9000–16,800	*Increase*
Switzerland	8000–12,000	*Increase*
Turkey	(14,000–20,000)	(Decrease)
Ukraine	18,000–28,800	Fluctuating
United Kingdom	92,000	*Decrease*

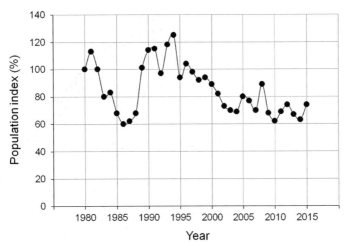

Figure 10.1 Change in % over the period 1980–2015 of the European population of common kestrel. Data collected from https://pecbms.info/trends-and-indicators/species-trends/species/f alco-tinnunculus/. Source of the data: EBCC/BirdLife/RSPB/CSO.

very large, and hence does not approach the thresholds for Vulnerable under the population size criterion (< 10,000 mature individuals with a continuing decline estimated to be > 10% in 10 years or three generations, or with a specified population structure).

The European population is estimated at 819,000–1,210,000 adult individuals (BirdLife International, 2015). Preliminary estimates place the world population size of the common kestrel in the range 4,000,000–6,500,000 adult individuals (Table 10.1). Although the species as a whole does not raise significant concerns, this might not be the case for small isolated populations or subspecies, such as those inhabiting islands.

10.4   Genetic Diversity

Genetic variation is a fundamental factor that drives the evolutionary adaptability of species to environmental changes. Genetic structure describes the total genetic diversity and its distribution within and among populations. Island populations often have low genetic diversity because of a reduced gene flow from mainland conspecific populations, small population size or past bottlenecks. Loss of genetic diversity and increased inbreeding (i.e. mating between individuals that are closely related genetically) can lead to the expression of deleterious recessive alleles, increased susceptibility to stressful environmental conditions in comparison to outbred individuals, or reduction in fecundity and reproductive success (*inbreeding depression*). Information on the genetic structure of a given population or species is therefore crucially important to develop effective conservation programmes.

10.4.1   The Special Case of the Mauritius Kestrel

The Mauritius kestrel is a renowned example of an island species that was on the brink of extinction. Prior to human settlement in 1750, the Mauritius kestrel was likely common on the island (McKelvey, 1977; Jones & Owadally, 1985). Habitat loss and degradation due to forest logging and agricultural expansion determined a rapid population decline, and by 1950 the species was rare and approaching extinction (Hachisuka, 1953). The increasing use of organochlorine pesticides further contributed to reduce the population size to one known breeding pair in 1974 (Safford & Jones, 1997).

A significant conservation effort ongoing since 1973 combining intensive captive-breeding and reintroduction programmes has enabled population restoration (Jones, 1987; Jones et al., 1995, 2013). It is, however, unclear what the peak population size was after the recovery and if there was further decline after this peak was achieved (Jones et al., 2013). By 2011–2012, the population was estimated to number around 300–400 individuals (Jones et al., 2013). Using microsatellite markers, a loss of genetic diversity and a reduction in effective population size (i.e. number of reproducing adults in the population) was found, consistent with the historical record of a severe bottleneck (Groombridge et al., 2000; Nichols et al., 2001). The ongoing climate changes make the future of the Mauritius kestrel very uncertain. Senapathi et al. (2011) showed that spring rainfall caused delays in the start of the breeding season, which might expose birds to risks of reduced reproductive success due to the adverse climatic conditions that occur later in the season. They concluded that these results, combined with the fact that the frequency of spring rainfall has increased by about 60% in the study area since 1962, imply that climate change might expose Mauritius kestrels to the stochastic risks of late reproduction. However, Mauritius kestrels have some capacity to respond to environmental changes by modifying their life-history strategies. Using 23 years of longitudinal data, Cartwright et al. (2014) found that females born in territories affected by anthropogenic habitat change (> 30% agricultural lands) started reproducing earlier in life than those in more natural territories. In so doing, they had both lower survival rates during the first two years of life and reproductive success later in life. These life-history responses resulted in an overall reproductive fitness similar to that of females breeding in more natural territories. Harvesting of eggs for artificial incubation and hand-rearing programmes also had an effect on individual life histories (Nicoll et al., 2006). Both males and females whose eggs were harvested (so causing a reduction in reproductive effort) had an increase in survival probability between the ages of 1 and 2 years. For a 1-year-old male kestrel the probability of reaching 5 years of age was 0.542 if he experienced egg harvesting and 0.296 if he did not (Nicoll et al., 2006). Similarly, for a 1-year-old female kestrel the probability of reaching 5 years of age was 0.575 if she experienced egg harvesting and 0.378 if she did not. This study indicated that species' recovery programmes might benefit from carefully considering the consequences of any conservation action on trade-offs among life-history traits (Nicoll et al., 2006).

10.4.2   The Seychelles Kestrel

The Seychelles kestrel provides another relevant example of an insular population bottle-neck, but in contrast to the Mauritius kestrel, it was able to recover with minimal conservation effort. Historical accounts described the Seychelles kestrel as common until the 1930s (Newton, 1867; Vesey-Fitzgerald, 1940). Subsequent field surveys suggested that the population underwent a crash and a subsequent recovery within the next 70 years (Vincent, 1966; Gaymer et al., 1969; Feare et al., 1974; Temple, 1977; Watson, 1981; Collar & Stuart, 1985; Kay et al., 2002). These observational data were corroborated by genetic analyses of both museum and living specimens. The analysis of microsatellites suggested that the global population of the Seychelles kestrel underwent a crash time between 1940 and the early 1970s, and at one time numbered as few as eight (estimated range: 3.5–22.0) individuals, which is compatible with the suggestion that there were fewer than 30 birds on Mahé during the 1960s (Groombridge et al., 2009).

10.4.3   Genetic Variation in the Common Kestrel

The examples of the Mauritius kestrel and of the Seychelles kestrel remind us that it is fundamental to monitor the status of insular kestrel species and even of insular subspecies in order to either avoid or buffer dramatic losses in genetic diversity. For example, the common kestrel occurs in the Cape Verde archipelago with two subspecies, *Falco tinnunculus neglectus* on the north-western islands and *F. t. alexandri* on the south-eastern islands. A study carried out over the period 1997–1999 found that both subspecies had lower allelic richness and heterozygosity than common kestrels living on the European or African mainland (Norway and Kenya in Hille, 2002; Austria in Hille et al., 2003). There was large among island variation in the observed heterozygosity, with the lowest and largest estimates of heterozygosity recorded on Santo Antão and on Santiago, respectively. Interestingly, whereas the heterozygosity on Santo Antão (0.22) was much lower than that recorded in Austria (0.56), the heterozygosity on Santiago (0.51) was close to that in Austria (Hille et al., 2003). The heterozygosity of Cape Verde kestrels was also lower than that recorded in kestrel subspecies inhabiting the Canary Islands (Table 10.3), probably because the Canary Islands are much closer to the mainland than the Cape Verde islands (about 100 km vs 450 km of Cape Verde). Hille et al. (2003) concluded that the genetic structure of common kestrels on Cape Verde islands was not affected by past severe bottlenecks, but was determined by patterns of migration from mainland to islands and from island to island. Thus, the geographic isolation of these island kestrel populations determined low immigration rates, which would explain the low levels of genetic diversity. Hill et al. (2003) also suggested that the north-eastern trade winds may have favoured dispersal of kestrels from north-east to south-west and, thus, would have significantly contributed to determine the spatial differences in genetic diversity. If so, the ongoing climatic changes might have an impact on the genetic structure of Cape Verde kestrels if the wind regimes will be modified.

The common kestrel also occurs on the Canary Islands with two subspecies, *F. t. canariensis* on Madeira, central and western islands, and *F. t. dacotiae* on eastern

Table 10.3 Observed heterozygosity (frequency of heterozygote individuals per locus) estimated from microsatellite analyses across several species, subspecies and conspecific populations of kestrel.

Species/subspecies	Locality	Observed heterozygosity	Study
F. araea	Seychelles	0.06	Nichols et al., 2001
F. naumanni		0.61	Nichols et al., 2001
F. newtoni	Madagascar	0.33	Nichols et al., 2001
F. punctatus (restored population)	Mauritius	0.10	Nichols et al., 2001
F. punctatus (ancestral population)	Mauritius	0.23	Nichols et al., 2001
F. rupicoloides		0.40	Nichols et al., 2001
F. rupicolus	South Africa	0.53	Nichols et al., 2001
F. tinnunculus alexandri	Brava	0.43	Hille et al., 2003
F. tinnunculus alexandri	Fogo	0.47	Hille et al., 2003
F. tinnunculus alexandri	Santiago	0.51	Hille et al., 2003
F. tinnunculus alexandri	Maio – Cape Verde	0.46	Hille et al., 2003
F. tinnunculus alexandri	Boavista – Cape Verde	0.28	Hille et al., 2003
F. tinnunculus alexandri	Sal – Cape Verde	0.30	Hille et al., 2003
F. tinnunculus canariensis	Canary Islands	0.58	Nichols et al., 2001
F. tinnunculus canariensis	La Palma – Canary Islands	0.40	Kangas et al., 2018
F. tinnunculus canariensis	Tenerife – Canary Islands	0.49	Kangas et al., 2018
F. tinnunculus canariensis	Gran Canaria – Canary Islands	0.55	Kangas et al., 2018
F. tinnunculus canariensis	La Gomera – Canary Islands	0.49	Kangas et al., 2018
F. tinnunculus canariensis	El Hierro – Canary Islands	0.43	Kangas et al., 2018
F. tinnunculus canariensis	Madeira – Canary Islands	0.53	Kangas et al., 2018
F. tinnunculus dacotiae	Fuerteventura – Canary Islands	0.58	Kangas et al., 2018
F. tinnunculus dacotiae	Lanzarote – Canary Islands	0.58	Kangas et al., 2018
F. tinnunculus neglectus	Santo Antão – Cape Verde	0.22	Hille et al., 2003
F. tinnunculus rufescens	Kenya	0.50	Nichols et al., 2001
F. tinnunculus tinnunculus	United Kingdom	0.61	Nichols et al., 2001
F. tinnunculus tinnunculus	Austria	0.56	Hille et al., 2003
F. tinnunculus tinnunculus	Mallorca	0.58	Kangas et al., 2018
F. tinnunculus tinnunculus	Ibiza	0.57	Kangas et al., 2018
F. tinnunculus tinnunculus	Menorca	0.55	Kangas et al., 2018
F. tinnunculus tinnunculus	Spain	0.64	Kangas et al., 2018
F. tinnunculus tinnunculus	Morocco	0.65	Kangas et al., 2018

islands, which differ in both morphology, colourations and genetics (Chapter 1). Both subspecies show low genetic variation and high inbreeding compared to continental subspecies (Alcaide et al., 2009; Kangas et al., 2018). Estimated population sizes are less than 9000 breeding pairs for *F. t. canariensis* and less than 500 breeding pairs for *F. t. dacotiae* (Carrillo, 2004, 2007; Madroño et al., 2004). For these reasons, *F. t. dacotiae* is currently classified as Vulnerable by the Spanish Ornithological Society/BirdLife (Madroño et al., 2004). Kangas et al. (2018) found evidence for past population bottlenecks in *F. t. dacotiae* and in nearby insular *F. t. tinnunculus* that caused decreased genetic diversity. Also, genetic diversity was lower in those populations living on the islands furthest from the continent (Table 10.3).

Spatial separation of kestrel populations inhabiting different islands therefore represents an important example of *isolation by distance* that leads to genetic differentiation. There are other processes that might generate genetic differentiation and that are, therefore, important to consider in population management strategies. One of these processes is linked to breeding asynchrony (i.e. individuals that breed consistently every year at different times of the season) that may cause *isolation by time* and thus genetic differentiation as suggested by one study on common kestrels in Italy (Casagrande et al., 2006a). Another process deals with a reduced gene flow between birds breeding in urban and rural environments (Rutkowski et al., 2006, 2010). Population fragmentation due to land-use changes may also reduce gene flow, which might jeopardise long-term persistence of populations due to inbreeding depression and loss of genetic diversity. Common kestrels show natal and breeding philopatry, which could reduce population connectivity further. However, kestrels also have dispersal capabilities that might counteract the effects of population fragmentation on the genetic diversity. Analyses of microsatellites did not reveal any significant genetic differentiation among European populations of common kestrels, but detected significant genetic differences in the more philopatric lesser kestrel (Alcaide et al., 2009). However, the genetic differentiation among lesser kestrel populations did not result in reduced genetic variation or increased inbreeding, indicating that severe population bottlenecks following extreme isolation might be needed to depauperate genetic variation (Alcaide et al., 2009).

10.5   Land-use Changes

Rapid and drastic land-use changes have led to substantial range contractions and extinctions of many bird species around the world. In Europe, agricultural intensification has led to landscape simplification/homogenisation with a significant loss in biodiversity as a consequence. Common kestrels appear to be particularly vulnerable to intensive agricultural practices (e.g. Shrubb, 1970, 1980; Schmid, 1990; Avilés et al., 2001; Schmid et al., 2001; Millon & Bretagnolle, 2004; Costantini et al., 2014; Nagy et al., 2017; Sumasgutner et al., 2019). In France, the common kestrel has experienced diminishing population size and contraction of its range since the 1990s (Millon & Bretagnolle, 2004). Butet et al. (2010) found that the highest abundance of kestrels occurs in French farming landscapes, where hedgerows, woods and grasslands are preserved. In contrast, a significant numerical contraction of kestrels occurs in those landscapes, where woodland habitats and meadows are highly reduced and where crops exceed 50% of the total land area (Butet et al., 2010). Although the use of artificial nest boxes may buffer the loss of natural breeding sites in agricultural landscapes, this might not be enough to sustain viable populations of kestrel because other factors may negatively affect their stability. A study carried out on kestrels breeding in nest boxes in the Po valley plain (Italy) found that pairs breeding in areas of intensive agriculture delayed egg-laying, had lower reproductive success and had offspring in poorer body condition compared with those breeding in temporary

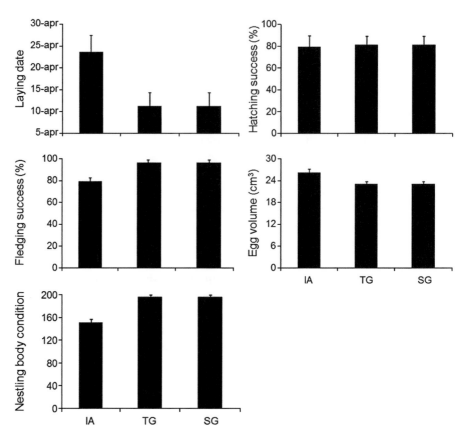

Figure 10.2 Kestrels breeding in areas characterised by intensive agricultural practices (IA) delayed egg-laying and had offspring in worse body condition as compared with kestrels breeding in temporary grasslands (TG) and permanent grasslands (PG). Hatching success, fledging success and egg volume were similar across land-use typologies. Least-square means and standard error are shown. Data of body condition are expressed as least square means of body mass (grams) obtained from linear mixed models with body size included as a covariate. See Costantini et al. (2014).

or permanent grasslands (Figure 10.2; Costantini et al., 2014). These results suggested that factors other than availability of breeding sites may have a significant impact on kestrels and should be given careful consideration when selecting the areas where to install nest boxes.

Shrubb suggested that intensive cereal cropping may affect the hunting behaviour of kestrels at critical times of year (Shrubb, 1970, 1980). Kestrels breeding in intensively cultivated areas might also face with higher energetic costs of foraging to reach good foraging patches compared with birds breeding in territories dominated by grasslands, which would translate in lower parental care. Higher energetic costs may also occur because kestrels would need to maintain larger home ranges in intensively cultivated areas (e.g. Village, 1990; Tella et al., 1998) in order to buffer a lower abundance of prey (Valkama et al., 1995) or a higher occurrence of smaller prey than in traditional agro-grazing systems (Tella et al., 1998). It is unlikely that the delayed breeding and lower nestling condition in intensively cultivated areas reported by Costantini et al. were due to lower-quality kestrels being forced to occupy intensively cultivated areas due to competition with high-quality kestrels because the body condition did not differ among adults breeding in different environments (Costantini et al., 2014). Thus, adults breeding in intensively cultivated areas might have actually limited their investment in parental care in favour of their self-maintenance.

It is clear that it is very important to improve land management strategies in a way that is compatible with the conservation of common kestrels or of other kestrel species breeding in agricultural landscapes, such as the lesser kestrel (Ursua et al., 2005; Catry et al., 2012). For example, rotation between cultivation and grasslands might limit the negative impact of intensive agriculture on the breeding activity of kestrels. Aschwanden et al. (2005) found that in intensively farmed regions a mosaic of different habitat types with low-intensity meadows and artificial grasslands (*ecological compensation areas*) mown at different times of the year together with undisturbed wild flower and herbaceous strips was best suited to provide a year-round supply of accessible prey for kestrels. Whittingham and Devereux (2008) showed that a reduction of grass height by mowing increased the number of foraging kestrels compared to control fields that were unmown. Garratt et al. (2011) found that breeding kestrels selected foraging habitats non-randomly, with cut grass (≤ 5 cm, all cut less than 2 weeks previously) being the most used relative to their availability because it likely provided better conditions for access to prey. Such a preference for fields after cutting declined by day 4 due to the rapid regrowth of grass (Peggie et al., 2011). Similarly, lesser kestrels also select given habitats for foraging or breeding. Ursua et al. (2005) showed that breeding lesser kestrels avoided irrigated maize and other crop types, while selecting for temporary grasslands of alfalfa. Catry et al. (2012) also found that allocating large fields of fallows around breeding colonies improved the reproductive success of lesser kestrels.

Habitat heterogeneity is, therefore, generally important for kestrels and has to be considered for any correct landscape planning and management. Timing and extent of habitat management are also important (e.g. mowing). It should also not be forgotten that the landscape variables that determine the distribution of common kestrels may differ between wintering and breeding seasons (e.g. Nantón, 2011) because birds have different requirements. Thus, longitudinal studies of kestrels' ecology and behaviour carried out over the whole year are fundamental to quantify any changes in habitat requirements adequately.

Agriculture intensification is not the only form of land use that threatens the conservation status of kestrels. A study on the changes in land use in Galicia (Spain) between 2001 and 2014 found a significant decline in the abundance of common kestrels and of other open-habitat specialist raptors (e.g. Montagu's harrier *Circus pygargus*) associated with increased intensive forest logging and a long-lasting trend of rural abandonment coupled with an unusually high frequency of wildfires and loss of mature heath and shrub formations (Tapia et al., 2017).

10.6   Conflicts with Man-made Structures

Fatal collisions of birds with human-made structures (e.g. electric power lines, wind turbines, windows) may represent a threat for bird conservation, particularly where such structures cross important corridors for migrating birds or resident species. Post-mortem analyses of 1483 carcasses of common kestrels at Monks Wood Experimental Station (England) collected between 1962 and 1997 found that collisions were probably the main human-related cause of death (Figure 10.3; Newton et al., 1999).

Collision of wildlife with wind turbines represents a growing concern worldwide. Barrios and Rodríguez (2004) quantified the bird mortality rate associated with two wind farms, including 87 wind turbines, over a period of one year in Spain. The mortality of common kestrels occurred during the annual peak of abundance in summer (mainly young) and was not associated with either the structural attributes of wind farms or visibility. The estimated mortality rate was 0.19 individuals per turbine per study year, which was the highest among the study species (Barrios & Rodríguez, 2004). Lower estimated mortality rates were recorded by

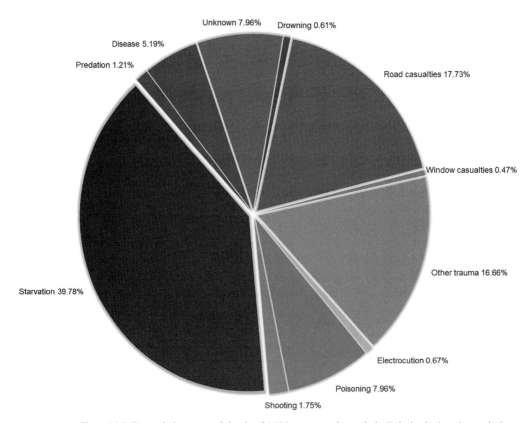

Figure 10.3 Recorded causes of death of 1483 common kestrels in Britain during the period 1962–97. Data collected from Newton et al. (1999).

Sebastián-González et al. (2018). They recorded 27 kestrel carcasses in Cádiz (869 turbines) from December 1993 to March 2016 and 515 kestrel carcasses in Castellón (320 turbines) from October 2006 to June 2015. In Portugal, Bernardino et al. (2012) found that the common kestrel was the most common raptor affected with an adjusted mortality of 0.011 fatalities for each turbine for each search performed. Morinha et al. (2014) found that 3.4% of the 59 carcasses found in 10 northern Portuguese wind farms between 2006 and 2011 were common kestrels. Wind turbines are also a threat for other kestrel species. It was estimated that between 73 and 333 American kestrels are killed annually by wind turbines at Altamont Pass Wind Resource Area in central California (Smallwood & Thelander, 2004).

Fatal collisions of kestrels with electric transmission lines appear of low concern globally as compared to other raptor species. This is not surprising because kestrels are agile, medium-sized birds that can easily avoid pylons or wires. However, collisions and electrocutions might be important causes of death at a local level, particularly in those areas rich in low- and medium-voltage overhead lines (e.g. Deschamps, 1980; Brochet, 1993; Bayle, 1999; Kabouche, 1999). For example, 130 carcasses of kestrels were found along 5 km of a 20-kV power line from 1988 to 1991 in France (Brochet, 1993). In south-west Spain, 14 electrocuted kestrels were found dead along about 14 km of power line from 1991 to 1993 (Janss, 2000). In Italy, a one-year search programme of carcasses of birds along about 79 km of power lines (132–380 kV) across seven localities did not record any kestrel carcasses (Costantini et al., 2017a). In another study carried out in Spain from January 2003 to February 2017 over an area including 188,751 poles recorded 17 kestrel electrocution casualties (Hernández-Lambraño et al., 2018). Nowadays, most power companies across Europe have enforced new protocols and criteria to reduce the conflict between birds and power lines: several dangerous lines have been retrofitted with bird-friendly solutions; new lines have been designed to reduce the risk of electrocution; and the wires have regularly been equipped with bird diverters. Another way to reduce collision and electrocution accidents has been through the burial of power lines. The efficacy of different forms of mitigation techniques (e.g. brush-type perch deflectors and rotating mirror perch deterrents on cross-arms, alternate designs of the pole-top insulator mount) was also tested experimentally, and specific designs vary according to manufacturer (e.g. APLIC, 2006; Ferrer, 2012; Dixon et al., 2019). The problem of electrocution persists, however, in less economically developed countries or in countries where power lines have been designed without taking into account the potential conflict with birds. For example, in Mongolia, 32 common kestrels were found dead along a three-phase 15-kV line covering 56 km over a period of one year (Dixon et al., 2019).

10.7   Top-down Limitation by Apex Avian Predators

There is increasing evidence that avian top predators can limit sympatric populations of small and medium-sized birds of prey through both direct predation or factors independent from predation (Terraube & Bretagnolle, 2018). Top avian predators may kill and eat both nestlings and adults of smaller birds of prey or affect their behaviour, distribution and reproductive success (Terraube & Bretagnolle, 2018). Decline in density or distribution of top predators can actually favour expansion of lower-rank predators through the mesopredator release effect (*mesopredator release hypothesis*; Soulé et al., 1988; Prugh et al., 2009). The impact of predation on population dynamics of common kestrels has been poorly explored so far. In northern England, Petty et al. (2003) found a negative relationship between abundances of common kestrels and goshawks and found that kestrels were 58.2% of the 239 raptors killed by goshawks during the study period (Figure 10.4). Importantly, most

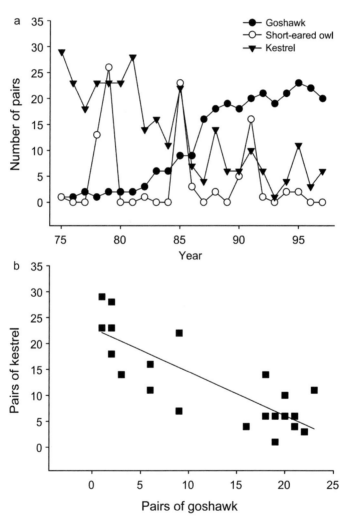

Figure 10.4 (a) Number of pairs of common kestrel, short-eared owl (diurnally active, vole-eating raptors) and northern goshawk in Kielder forest during the period 1975–1997. (b) Relationship between number of pairs of common kestrels and northern goshawks ($r^2 = 0.70$, $p < 0.001$). Each point represents one year during the period 1975–1997. Reprinted with slight modifications from Petty et al. (2003) with permission from John Wiley and Sons.

kestrels were adults and were predated in early spring, prior to breeding, in an area that could contain a limited number of pairs. These data suggested that goshawks might have had the potential to remove all breeders and, possibly, all non-breeder kestrels passing throughout the study site. Similarly, prior work on American kestrels suggested that predation by Cooper's hawks (*Accipiter cooperii*) might be relevant (Farmer et al., 2006).

Although these data did not provide insight on whether predation on kestrels is important and regulates population dynamics, they suggested that plans aimed at restoring populations of large raptor species might come at a cost for smaller raptors, such as kestrels. These costs might also emerge because of competition for the same resources, such as food or breeding sites, as shown in both nocturnal and diurnal birds of prey (Kostrzewa, 1991; Hakkarainen & Korpimäki, 1996).

Predation by lower-level predators may also be relevant. Fargallo et al. (2001) identified that rats (*Rattus norvegicus*) and dormice (*Elyomis quercinus*) were important predators of eggs in nests located in buildings. In nest boxes or nests located on trees, ravens, carrion crows, European genets (*Genetta genetta*) and stone martens (*Martes foina*) were seen frequently to predate eggs, nestlings or incubating adults (Fargallo et al., 2001).

10.8   Population Composition and Dynamics

In an increasingly human-modified world, efficient conservation plans require a detailed understanding of population dynamics. Territorial behaviour, competition and availability of both food and nest sites are important drivers of population composition and turnover in birds of prey. Little is still known about the role of these factors in regulating populations of the common kestrel across all its distribution area.

Population turnover is one important feature of the population dynamics. It refers to the demographic changes of a given population, i.e. the recruitment and loss of individuals over time due to birth or immigration and death or emigration (Village, 1990). These demographic data are important to estimate the population growth rate. One way to determine population turnover is by marking individual birds and following their individual history. Mark and recapture also enables estimation of the lifespan of individuals. European longevity records recorded by European bird ringing indicate that the two oldest kestrels marked so far were 20 years and 5 months old (found sick) and 16 years and 5 months old (found dead; Fransson et al., 2017).

Village (1985b) found that greater numbers of kestrels in good vole years outside the breeding season were due to more juveniles settling in autumn and persisting longer through the winter. There are clearly several factors that affect the wintering population composition. There are both year-round and summer-only residents. The choice to stay or migrate might depend on individual age (i.e. prior experience) or genetics. Reduced availability of food might also stimulate a cohort of individuals to leave the area. There are also individuals that come from other populations. Incomers are usually juveniles, which disperse after they become fully independent from their parents. There is also a surplus of non-breeding individuals (mainly first-year birds) capable of mating with breeding birds that lost their partners (e.g. by experimental removal of one pair member; Village, 1990) or of occupying previously unavailable nesting sites (e.g. provisioning of nest boxes late in the season; Village, 1990). The role of surplus individuals in regulating population dynamics is currently poorly understood.

Mortality is another important factor during wintertime. Death of individuals may reduce the population size below the carrying capacity of the environment, which could explain why a rapid increase in population size in early spring has been observed in some populations (Village, 1982b). Early work carried out at high latitudes showed that about 30–40% of young might survive until the next breeding season, while survival of adults is around twice that of young (Village, 1990). Moreover, survival of young seemed to increase in good vole years and mild winters, and was generally higher in early- than in late-fledged individuals (Village, 1990). In many European countries, monitoring programmes are operated to collect demographic data on a wide range of breeding bird species. Using a long-term data set gathered by volunteers across Great Britain, Robinson et al. (2014) estimated that the survival probability of common kestrels was lower over the immediate post-fledging period (mean and 95% CI: 0.29, 0.14–0.51) compared to that of young until fledging (0.91, 0.79–0.98) or at adulthood (0.66, 0.53–0.80). In Switzerland, Fay et al. (2019) found that the estimated average annual survival probabilities in kestrels breeding in nest boxes were 0.72 for adults and 0.49 for young using a database of 15 years. Mortality of young kestrels was also found to be particularly high during the post-fledging dependence period in other studies. Bustamante (1994) estimated a mortality of 9% in Spain, while Boileau and Bretagnolle (2014) estimated a mortality of 12% in France. Thus, mortality of kestrels is particularly high during the post-fledging life until their first spring, meaning that selective disappearance of specific cohorts of individuals might play a key role in structuring the phenotypic and genetic variation of the adult population.

Village (1990) also observed that year-round residents had a mean stay in the area of around 2 years. This estimate of residence duration was similar for males and females, suggesting they had fairly similar local survival rates. However, Village (1990) also observed that prior poor breeding performance (e.g. reproductive failure) was less strongly associated with subsequent return of females than of males (divorce behaviour in Chapter 7). In other words, Village (1990) found that females were more likely to breed with their partner, if still present in the area, the following year if their previous reproductive attempt was successful. If not, females more than males tended to change the partner and even abandon the area, resulting in higher individual turnover. Vasko et al. (2011) also found a high annual turnover rate in Finland (about 90% and 85% of female and male parents are new each year, respectively), with a high divorce rate of females. Territoriality might also be important in determining the probability of an individual leaving or entering a given area. Common kestrels are territorial. It is, however, unknown the degree to which aggression towards incomers contributes to regulate population size, density and composition. This might be relevant because territory size decreases with increased population density (Village, 1990).

Availability of prey is one important driver of population size, composition and dynamics. Kestrels show two general responses to fluctuations in food availability. In Europe, populations at high latitudes, which rely mostly on a single prey, show demographic responses (e.g. natality, mortality) or adjust their movement ecology (e.g. migration) to the strong cyclic population dynamics that characterise vole communities compared to southern regions (e.g. Hanski et al., 2001; Korpimäki et al., 2005; Fargallo et al., 2009). At lower latitudes, kestrels are more food-generalists, so that they may shift to alternate prey or increase competition for food whether prey fluctuations occur. In Scotland, Village (1990) found that, when voles were scarce, there were fewer kestrels in the study area and most were adults, implying higher mortality or emigration of young. In contrast, when voles were abundant, the number of kestrels was greater and most were first-year birds. A study carried out in Finland from 1977 to 1987 found that annual changes in the number of breeding birds were synchronous with those in the estimated density of *Microtus* voles (Figure 10.5; Korpimäki & Norrdahl, 1991). The number of *Microtus* voles eaten was linearly related to their density in the study area (Korpimäki & Norrdahl, 1991). Changes in immigration and emigration were two important responses of kestrels to *Microtus* fluctuations. In contrast, competition for nest sites was unlikely to be significant because vacant nest sites were available even in the peak vole years (Korpimäki & Norrdahl, 1991).

Further work showed that the spring densities of kestrels were not related to vole abundance in the preceding autumn, supporting the lack of a time lag in the response of kestrels to fluctuations in population size of their main prey (Korpimäki, 1994). In southern Sweden, the response of kestrels to prey populations was mainly functional (change in feeding habits) rather than numerical, and a time delay in the number of voles eaten compared to their local abundance was observed (Erlinge et al., 1983). A significant connection between kestrel and vole population dynamics has also been described at the southern latitudes of Spain. Fargallo et al. (2009) found that the growth rate of a kestrel population in the Campo Azálvaro region was higher in those years when common voles (*Microtus arvalis*) were more abundant.

The collection of high-quality demographic data is also required to identify source and sink populations, which is a major step for preparing a conservation plan and making good management decisions (Hernández-Matías et al., 2013; Loreau et al., 2013). In source populations, births exceed deaths, thus a quota of individuals migrate in sink populations, where deaths exceed births (Pulliam, 1988; Loreau et al., 2013). Thus, a source population has the potential to maintain several sink populations as long as these are not ecological traps, which could jeopardise the conservation status of the species (Delibes et al, 2001). Reproductive data, age structure, dispersal and survival are key demographic data to collect. Regarding survival, estimating the true survival is challenging, so apparent survival is often used. However, apparent survival is confounded with permanent emigration, meaning that a kestrel that is no longer seen at the

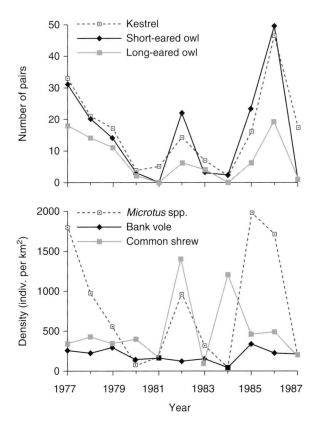

Figure 10.5 (a) Variation in the number of kestrel, short-eared owl and long-eared owl pairs breeding at Alajoki during the period 1977–1987. (b) Variation in the estimated spring density of *Microtus* voles, bank voles, and common shrews at Alajoki during the period 1977–1987. Reprinted with slight modifications from Korpimäki and Norrdahl (1991) with permission from John Wiley and Sons.

study area might simply have left it. Relying on apparent survival may thus underestimate the population growth rate.

10.9   Conclusions

Protective legislation has played a significant role in either buffering population decline or favouring population recovery (e.g. Noer & Secher, 1983). The common kestrel is thus now evaluated as Least Concern at global level. However, at the European level, the species is considered of conservation concern owing to a continuous moderate decline since the 1980s. Agriculture intensification in Europe with consequent reduction in good foraging areas and in availability of nest sites has been the main threat for the conservation of common kestrels. Although collisions with power lines or wind turbines appear to be of low concern, it should not be forgotten that multiple threats could combine additively to affect the viability of populations. These factors are very important to consider by stakeholders and policy makers if we are to develop evidence-based landscape planning to limit anthropogenic impacts on wild-life. In this context, agri-environment measures of the European Community are a key example for the integration of environmental concerns into the Common Agricultural Policy. They are designed to encourage farmers to protect and enhance the environment on their farmland by paying them for the provision of environmental services. Agri-environment measures may be designed at the national, regional or local level, so that they can be adapted to particular farming systems and specific environmental conditions.

Our poor knowledge of the demography and population dynamics of the common kestrel across its entire distribution has greatly limited our capacity to counteract the impact of anthropogenic activities on the species. The great recovery from the brink of extinction of the Mauritius kestrel, which is regarded as one of the most remarkable raptor conservation programmes in the world, teaches us that management of species of conservation concern would greatly benefit from consideration of life-history theory. It is also very important to quantify the impact of predation, but also not to lose sight of the fact that kestrels are predators too and may, therefore, have a negative impact on the conservation status of other species (Smart & Amar, 2018). Thus, management of kestrel populations and reintroduction programmes have to carefully consider all the pros and cons for the kestrel and for other species of conservation concern.

References

Acosta, I., Hernández, S., Gutiérrez, P. N., et al. (2010). Acuaroid nematodes in the common kestrel (*Falco tinnunculus*) in the south of Spain. *Vet. J.*, 183, 234–7.

Adriaensen, F., Verwimp, N. and Dhondt, A. A. (1997). Are Belgian kestrels *Falco tinnunculus* migratory: an analysis of ringing recoveries. *Ring. Migrat.*, 18, 91–101.

Adriaensen, F., Verwimp, N. and Dhondt, A. A. (1998). Between cohort variation in dispersal distance in the European kestrel *Falco tinnunculus* as shown by ring recoveries. *Ardea*, 86, 147–52.

Afonso, S., Vanore, G. and Batlle, A. (1999). Protoporphyrin IX and oxidative stress. *Free Radic. Res.*, 31, 161–70.

Agarwal, A. and Allamaneni, S.S. (2004). Role of free radicals in female reproductive diseases and assisted reproduction. *Reprod. Biomed. Online*, 9, 338–47.

Albers, P. H., Koterba, M. T., Rossmann, R., et al. (2007). Effects of methylmercury on reproduction in American kestrels. *Environ. Toxicol. Chem.*, 26, 1856–66.

Alcaide, M., Negro, J. J., Serrano, D., Tella, J. L. and Rodríguez, C. (2005). Extra-pair paternity in the Lesser Kestrel *Falco naumanni*: a re-evaluation using microsatellite markers. *Ibis*, 147, 608–11.

Alcaide, M., Serrano, D., Negro, J. J., et al. (2009). Population fragmentation leads to isolation by distance but not genetic impoverishment in the philopatric lesser kestrel: a comparison with the widespread and sympatric Eurasian Kestrel. *Heredity*, 102, 190–8.

Alexander, B. (1898). Further notes on the ornithology of the Cape Verde Islands. *Ibis*, 277–85.

Allen, T. and Clarke, J. A. (2005). Social learning of food preferences by white-tailed ptarmigan chicks. *Anim. Behav.*, 70, 305–10.

Alonso-Alvarez, C., Bertrand, S., Faivre, B., Chastel, O. and Sorci, G. (2007). Testosterone and oxidative stress: the oxidation handicap hypothesis. *Proc. R. Soc. Lond. B*, 274, 819–25.

Amundsen, T. (2000). Why are female birds ornamented? *Trends Ecol. Evol.*, 15, 149–55.

Andersson, M. (1994). *Sexual selection*. Princeton, NJ: Princeton University Press.

Antoniazza, S., Burri, R., Fumagalli, L., Goudet, J. and Roulin, A. (2010). Local adaptation maintains clinal variation in melanin-based coloration of European barn owls (*Tyto alba*). *Evolution*, 64, 1944–54.

Anushiravani, S. and Roshan, Z. S. (2017). Identification of the breeding season diet of the Common Kestrel, *Falco tinnunculus* in the north of Iran. *Zool. Ecol.*, 27, 114–6.

Aparicio, J. M. (1994a). The effect of variation in the laying interval on proximate determination of clutch size in the European Kestrel. *J. Avian Biol.*, 25, 275–80.

Aparicio, J. M. (1994b). The seasonal decline in clutch size: an experiment with supplementary food in the kestrel. *Oikos*, 71, 451–8.

Aparicio, J. M. (1998). Individual optimization may explain differences in breeding time in the European kestrel *Falco tinnunculus*. *J. Avian Biol.*, 29, 121–8.

Aparicio, J. M. (1999). Intraclutch egg-size variation in the Eurasian kestrel: advantages and disadvantages of hatching from large eggs. *The Auk*, 116, 825–30.

Aparicio, J. M. (2000). Differences in the diets of resident and non-resident kestrels in Spain. *Ornis Fenn.*, 77, 169–75.

Aparicio, J. M. and Bonal, R. (2002). Effects of food supplementation and habitat selection on timing of lesser kestrel breeding. *Ecology*, 83, 873–7.

Avian Power Line Interaction Committee (APLIC) (2006). *Suggested practices for avian protection on power lines: the state of the art in 2006*. Washington, DC: Edison Electric Institute.

Ardia, D. R. (2006). Glycated hemoglobin and albumin reflect nestling growth and condition in American kestrels. *Comp. Biochem. Physiol. Part A*, 143, 62–6.

Ardia, D. R. and Bildstein, K. L. (1997). Sex-related differences in habitat selection in wintering American kestrels, *Falco sparverius. Anim. Behav.*, 53, 1305–11.

Aschwanden, J., Birrer, S. and Jenni, L. (2005). Are ecological compensation areas attractive hunting sites for common kestrels (*Falco tinnunculus*) and long-eared owls (*Asio otus*)? *J. Ornithol.*, 146, 279–86.

Ashmole, N. P. (1961). *The biology of certain terns*. PhD thesis, University of Oxford, Oxford, UK.

Aumann, T. (1988). The diet of the brown goshawk, *Accipiter fasciatus*, in south-eastern Australia. *Aust. Wildl. Res.*, 15, 587–94.

Avilés, J. M., Sánchez, J. M. and Parejo, D. (2001). Breeding rates of Eurasian Kestrels *Falco tinnunculus* in relation to surrounding habitat in southwest Spain. *J. Raptor Res.*, 35, 31–4.

Balfour, E. (1955). Kestrels nesting on the ground in Orkney. *Bird Notes*, 26, 245–53.

Balgooyen, T. (1976). Behaviour and ecology of the American kestrel (*Falco sparverius*) in the Sierra Nevada of California. *Univ. Calif. Publ. Zool.*, 103, 1–83.

Barnard, P. (1987). Foraging site selection by three raptors in relation to grassland burning in a montane habitat. *Afr. J. Ecol.*, 25, 35–45.

Barrios, L. and Rodríguez, A. (2004). Behavioural and environmental correlates of soaring-bird mortality at on-shore wind turbines. *J. Appl. Ecol.*, 41, 72–81.

Bauer, H. G., Bezzel, E. and Fiedler, W. (2005). *Das kompendium der vögel mitteleuropas* (Bd. 1: Nonpasseriformes). Wiebelsheim: Aula-Verlag.

Bautista, L. M., Alonso, J. C. and Alonso, J. (1995). A field test of ideal distribution of in flock feeding common cranes. *J. Anim. Ecol.*, 64, 747–57.

Bautista, L. M., Alonso, J. C. and Alonso, J. (1998). Foraging site displacement in common crane flocks. *Anim. Behav.*, 56, 1237–43.

Bayle, P. (1999). Preventing birds of prey problems at transmission lines in Western Europe. *J. Raptor Res.*, 33, 43–8.

Baziz, B., Souttu, K., Doumandji, S. and Denys, C. (2001). Quelques aspects sur le régime alimentaire du faucon crécerelle *Falco tinnunculus* (Aves, Falconidae) en Algérie. *Alauda*, 69, 413–8.

Beamonte-Barrientos, R., Velando, A. and Torres, R. (2014). Age-dependent effects of carotenoids on sexual ornaments and reproductive performance of a long-lived seabird. *Behav. Ecol. Sociobiol.*, 68, 115–26.

Becker, J. J. (1987). Revision of '*Falco*' *ramenta* Wetmore and the Neogene evolution of the Falconidae. *The Auk*, 104, 270–6.

Beichle, U. (1980). Siedlungsdichte, jagdreviere und jagdweise des turmfalken (*Falco tinnunculus*) im stadtgebiet von Kiel. *Corax*, 8, 3–12.

Benathan, M., Virador, V., Furumura, M., et al. (1999). Coregulation of melanin precursors and tyrosinase in human pigment cells: roles of cysteine and glutathione. *Cell Mol. Biol.*, 45, 981–90.

Ben-David, M. and Flaherty, E. A. (2012). Stable isotope in mammalian research: a beginner's guide. *J. Mammal.*, 93, 312–28.

Benson, C. W. (1967). The birds of Aldabra and their status. *Atoll Res. Bull.*, 118, 63–111.

Benson, C. W. and Penny, M. J. (1971). The land birds of Aldabra. *Phil. Trans. R. Soc. Lond. B*, 206, 417–527.

Bergier, P. (1987). *Les rapaces diurnes du Maroc. Statut, répartition et ecologie.* Aix-en-Provence: Annales du C.E.E.P. (ex-C.R.O.P.) nr. 3.

Bernardino, J., Zina, H., Passos, I., et al. (2012). Bird and bat mortality at Portuguese wind farms. In *IAIA12 Conference Proceedings Energy Future: the Role of Impact Assessment.* 32nd Annual Meeting of the International Association for Impact Assessment, 27 May–1 June 2012, Porto, Portugal (pp. 1–5). Fargo, ND: IAIA.

Beukeboom, L., Dijkstra, C., Daan, S. and Meijer, T. (1988). Seasonality of clutch size determination in the kestrel *Falco tinnunculus*: an experimental approach. *Ornis Scandinav.*, 19, 41–8.

Bird, D. M. and Laguë, P. C. (1982a). Fertility, egg weight loss, hatchability, and fledging success in replacement clutches of captive American kestrels. *Can. J. Zool.*, 60, 80–8.

Bird, D. M. and Laguë, P. C. (1982b). Influence of forced renesting, seasonal date of laying, and female characteristics on clutch size and egg traits in captive American kestrels. *Can. J. Zool.*, 60, 71–9.

Bird, D. M., Weil, P. G. and Lague, P. C. (1980). Photoperiodic induction of multiple breeding seasons in captive American kestrels. *Can. J. Zool.*, 58, 1022–6.

BirdLife International (2015). *European Red List of birds.* Luxembourg: Office for Official Publications of the European Communities.

BirdLife International (2016a). *Falco alopex.* The IUCN Red List of Threatened Species 2016: e.T22696402A93559888.

BirdLife International (2016b). *Falco araeus.* The IUCN Red List of Threatened Species 2016: e.T22696380A93558237.

BirdLife International (2016c). *Falco ardosiaceus.* The IUCN Red List of Threatened Species 2016: e.T22696406A93560247.

BirdLife International (2016d). *Falco cenchroides.* The IUCN Red List of Threatened Species 2016: e.T22696391A93558789.

BirdLife International (2016e). *Falco dickinsoni.* The IUCN Red List of Threatened Species 2016: e.T22696410A93560617.

BirdLife International (2016f). *Falco moluccensis.* The IUCN Red List of Threatened Species 2016: e.T22696388A93558606.

BirdLife International (2016g). *Falco naumanni.* The IUCN Red List of Threatened Species 2016: e.T22696357A87325202.

BirdLife International (2016h). *Falco newtoni.* The IUCN Red List of Threatened Species 2016: e.T22696368A93557702.

BirdLife International (2016i). *Falco punctatus.* The IUCN Red List of Threatened Species 2016: e.T22696373A93557909.

BirdLife International (2016j). *Falco rupicoloides.* The IUCN Red List of Threatened Species 2016: e.T22696398A93559628.

BirdLife International (2016k). *Falco sparverius.* The IUCN Red List of Threatened Species 2016: e.T22696395A93559037.

BirdLife International (2016l). *Falco tinnunculus*. The IUCN Red List of Threatened Species 2016: e.T22696362A93556429.

BirdLife International (2016m). *Falco zoniventris*. The IUCN Red List of Threatened Species 2016: e.T22696414A93560862.

BirdLife International (2017). *European birds of conservation concern: populations, trends and national responsibilities*. Cambridge, UK: BirdLife International.

BirdLife International and Handbook of the Birds of the World (2017). *Bird species distribution maps of the world*. Version 7.0. Available at http://datazone.birdlife.org/species/requestdis.

Blanckenhorn, W. U. (2000). The evolution of body size: what keeps organisms small? *Q. Rev. Biol.*, 75, 385–407.

Blanco, G., Martínez-Padilla, J., Dávila, J. A., Serrano, D. and Viñuela, J. (2003a). First evidence of sex differences in the duration of avian embryonic period: consequences for sibling competition in sexually dimorphic birds. *Behav. Ecol.*, 14, 702–6.

Blanco, G., Martínez-Padilla, J., Serrano, D., Dávila, J. A. and Viñuela, J. (2003b). Mass provisioning to different-sex eggs within the laying sequence: consequences for adjustment of reproductive effort in a sexually dimorphic bird. *J. Anim. Ecol.*, 72, 831–8.

Blas, J., Perez-Rodriguez, L., Bortolotti, G. R., Vinuela, J. and Marchant, T. A. (2006). Testosterone increases bioavailability of carotenoids: insights into the honesty of sexual signalling. *Proc. Natl Acad. Sci. USA*, 103, 18633–7.

Blévin, P., Angelier, F., Tartu, S., et al. (2017). Perfluorinated substances and telomeres in an Arctic seabird: cross-sectional and longitudinal approaches. *Environ. Pollut.*, 230, 360–7.

Boev, Z. (1999). *Falco bakalovi* sp. n. – a Late Pliocene falcon (Falconidae, Aves) from Varshets (W Bulgaria). *Geol. Balcan.*, 29, 131–5.

Boev, Z. (2011a). *Falco bulgaricus* sp. n. (Aves: Falconiformes) from the Late Miocene of Hadzhidimovo (SW Bulgaria). *Acta Zool. Bulg.*, 63, 17–35.

Boev, Z. (2011b). New fossil record of the Late Pliocene kestrel (*Falco bakalovi* Boev, 1999) from the type locality in Bulgaria. *Geol. Balcan.*, 40, 13–30.

Boev, Z. (2011c). *Falco bulgaricus* sp. n. (Aves: Falconiformes) from the Late Miocene of Hadzhidimovo (SW Bulgaria). *Acta Zool. Bulg.*, 63, 17–35.

Boileau, N. and Bretagnolle, V. (2014). Post-fleding dependence period in the Eurasian kestrel (*Falco tinnunculus*) in western Spain. *J. Raptor. Res.*, 48, 248–56.

Boileau, N., Delelis, N. and Hoede, C. (2006). Utilisation de l'espace et de l'habitat par le Faucon crécerelle *Falco tinnunculus* en période de reproduction. *Alauda*, 74, 251–64.

Bondurianskyi, R. and Day, T. (2009). Nongenetic inheritance and its evolutionary implications. *Annu. Rev. Ecol. Evol. System*, 40, 103–25.

Bonin, B. and Strenna, L. (1986). Sur la biologie du Faucon crécerelle *Falco tinnunculus* en Auxois. *Alauda*, 54, 241–62.

Bortolotti, G. R. and Iko, W. M. (1992). Non-random pairing in American kestrels: mate choice versus intra-sexual competition. *Anim. Behav.*, 44, 811–21.

Bortolotti, G. R. and Wiebe, K. L. (1993). Incubation behaviour and hatching patterns in the American Kestrel *Falco sparverius*. *Ornis Scand.*, 24, 41–7.

Bortolotti, G. R., Negro, J. J., Tella, J. L., Marchant, T. A. and Bord, D. (1996). Sexual dichromatism in birds independent of diet, parasites and androgens. *Proc. R. Soc. Lond. B*, 263, 1171–6.

Bortolotti, G. R., Tella, J. L., Forero, M. G., Dawson, R. D. and Negro, J. J. (2000). Genetics, local environment and health as factors influencing serum carotenoids in wild American kestrels (*F. sparverius*). *Proc. R. Soc. Lond. B*, 267, 1433–8.

Bortolotti, G. R., Fernie, K. J. and Smits, J. E. (2003). Carotenoid concentration and coloration of American kestrels (*Falco sparverius*) disrupted by experimental exposure to PCBs. *Funct. Ecol.*, 17, 651–7.

Bounas, A., Tsaparis, D., Gustin, M., et al. (2018). Using genetic markers to unravel the origin of birds converging towards pre-migratory sites. *Scient. Rep.*, 8, 8326.

Bourne, W. R. P. (1955). A new race of kestrel *Falco tinnunculus* Linnaeus from the Cape Verde Islands. *Bull. Brit. Ornith. Club*, 75, 35–6.

Boyce, D. A. and White, C. W. (1987). Evolutionary aspects of kestrel systematics: a scenario. In D. M. Bird and R. Bowman, eds., *The ancestral kestrel* (pp. 1–21). Quebec: Raptor Res. Found., Inc.

Brichetti, P. and Fracasso, G. (2003). *Ornitologia italiana*. Vol. 1. Bologna: Alberto Perdisa.

Brochet, J. (1993). *Expérimentation de prototypes: spirale (SAAE) et piver (RAYCHEM) sur les lignes EDF MT 20 000 volts Compertrix – Haussimont (1991–92)*. Rochefort, France: LPO and EDF.

Brockmann, H. J. and Barnard, C. J. (1979). Kleptoparasitism in birds. *Anim. Behav.*, 27, 487–514.

Brommer, J. E., Wilson, A. J. and Gustafsson, L. (2007). Exploring the genetics of aging in a wild passerine bird. *Am. Nat,* 170, 643–50.

Brown, L. H. and Amadon, D. (1968). *Eagles, hawks and falcons of the world*. London: Country Life Books.

Brown, R. G. B. (1969). Seed selection of pigeons. *Behaviour*, 34, 115–31.

Bryan, J. R. (1984). Factors influencing differential predation on house mice (*Mus musculus*) by American kestrels (*Falco sparverius*). *Raptor Res.*, 18, 143–7.

Bush, N. G., Brown, M. and Downs, C. T. (2008). Seasonal effects on thermoregulatory responses of the Rock Kestrel, *Falco rupicolis*. *J. Therm. Biol.*, 33, 404–12.

Bustamante, J. (1994). Behavior of colonial common kestrels (*Falco tinnunculus*) during the post-fledging dependence period in Southwestern Spain. *J. Rapt. Res.*, 28, 79–83.

Bustamante, J. and Negro, J. J. (1994). The post-fledging dependence period of the Lesser Kestrel (*Falco naumanni*) in southwestern Spain. *J. Rapt. Res.*, 28, 158–63.

Bustnes, J. O., Erikstad, K. E., Lorentsen, S. H. and Herzke, D. (2008). Perfluorinated and chlorinated pollutants as predictors of demographic parameters in an endangered seabird. *Environ. Pollut.*, 156, 414–24.

Butet, A., Michel, N., Rantier, Y., et al. (2010). Responses of common buzzard (*Buteo buteo*) and Eurasian kestrel (*Falco tinnunculus*) to land use changes in agricultural landscapes of Western France. *Agricult. Ecosyst. Environm.*, 138, 152–9.

Byrne, M., Sewell, M. A. and Prowse, T. A. A. (2008). Nutritional ecology of sea urchin larvae: influence of endogenous and exogenous nutrition on echinopluteal growth and phenotypic plasticity in Tripneustes gratilla. *Funct. Ecol.*, 22, 643–8.

Cade, T. J. and Digby, D. R. (1982). *The falcons of the world*. New York: Cornell University Press.

Campbell, B. and Lack, E. (1985). *A dictionary of birds*. London: British Ornithologists' Union.

Campbell, R. W. (1985). First record of the Eurasian kestrel for Canada. *Condor*, 87, 294.

Candolin, U. (2003). The use of multiple cues in mate choice. *Biol. Rev.*, 78, 575–95.

Canestrelli, D., Bisconti, R. and Carere, C. (2016). Bolder takes all? The behavioral dimension of biogeography. *Trends Ecol. Evol.*, 31, 35–43.

Carere, C. and Maestripieri, D. (2013). *Animal personalities: behavior, physiology, and evolution*. Chicago: The University of Chicago Press.

Carrillo, J. (2004). Cernícalo vulgar (*Falco tinnunculus dacotiae*). In A. Madroño, C. González and J. C. Atienza, eds., *Libro rojo de las aves de España* (pp. 164–6). Madrid: Dirección general da biodiversidad/Ministerio de Medio Ambiente–SEO/BirdLife.

Carrillo, J. (2007). Cernícalo vulgar, *Falco tinnunculus*. In J. A. Lorenzo, ed., *Atlas de las aves nidificantes en el archipiélago canario (1997–2003)* (pp. 173–8). Madrid: Dirección General de Conservación de la Naturaleza–Sociedad Española de Ornitología.

Carrillo, J. and Aparicio, J. M. (2001). Nest defence behaviour of the Eurasian kestrel (*Falco tinnunculus*) against human predators. *Ethology*, 107, 865–75.

Carrillo, J. and González-Dávila, E. (2005). Breeding biology and nest characteristics of the Eurasian Kestrel in different environments on an Atlantic island. *Ornis Fenn.*, 82, 55–62.

Carrillo, J. and González-Dávila, E. (2009). Latitudinal variation in breeding parameters of the common kestrel *Falco tinnunculus*. *Ardeola*, 56, 215–28.

Carrillo, J. and González-Dávila, E. (2010a). Geo-environmental influences on breeding parameters of the Eurasian kestrel (*Falco tinnunculus*) in the Western Palaearctic. *Ornis Fenn.*, 87, 15–25.

Carrillo, J. and González-Dávila, E. (2010b). Impact of weather on breeding success of the Eurasian kestrel *Falco tinnunculus* in a semi-arid island habitat. *Ardea*, 98, 51–8.

Carrillo, J. and González-Dávila, E. (2013). Aggressive behaviour and nest-site defence during the breeding season in an island kestrel population. *J. Ethol.*, 31, 211–8.

Carrillo, J., González-Dávila, E. and Ruiz, X. (2017). Breeding diet of Eurasian Kestrels *Falco tinnunculus* on the oceanic island of Tenerife. *Ardea*, 105, 99–111.

Carson, R. (1962). *Silent spring*. Boston, MA: Houghton Mifflin.

Cartwright, S. J., Nicoll, M. A. C., Jones, C. G., Tatayah, V. and Norris, K. (2014). Anthropogenic natal environmental effects on life histories in a wild bird population. *Curr. Biol.*, 24, 536–40.

Casagrande, S., Dell'Omo, G., Costantini, D. and Tagliavini, J. (2006a). Genetic differences between early- and late-breeding Eurasian kestrels. *Evol. Ecol Res.*, 8, 1029–38.

Casagrande, S., Csermely, D., Pini, E., Bertacche, V. and Tagliavini, J. (2006b). Skin carotenoid concentration correlates with male hunting skill and territory quality in the kestrel (*Falco tinnunculus*). *J. Avian Biol.*, 37, 190–6.

Casagrande, S., Costantini, D., Fanfani, A., Tagliavini, J. and Dell'Omo, G. (2007). Patterns of serum carotenoid accumulation and skin color variation in nestling kestrels in relation to breeding conditions and different terms of carotenoid supplementation. *J. Comp. Physiol. B*, 177, 237–45.

Casagrande, S., Nieder, L., Di Minin, E., La Fata, I. and Csermely, D. (2008). Habitat utilization and prey selection of the kestrel *Falco tinnunculus* in relation to small mammal abundance. *Ital. J. Zool.*, 75, 401–9.

Casagrande, S., Costantini, D., Tagliavini, J. and Dell'Omo, G. (2009). Phenotypic, genetic, and environmental causes of variation in yellow skin pigmentation and serum carotenoids in Eurasian kestrel nestlings. *Ecol. Res.*, 24, 273–9.

Casagrande, S., Dell'Omo, G., Costantini, D., Tagliavini, J. and Groothuis, T. (2011). Variation of a carotenoid-based trait in relation to oxidative stress and endocrine status during the breeding season in the Eurasian kestrel: a multi-factorial study. *Comp. Biochem. Physiol. Part A*, 160, 16–26.

Casagrande, S., Costantini, D., Dell'Omo, G., Tagliavini, J. and Groothuis, T. G. G. (2012). Differential effects of testosterone metabolites oestradiol and dihydrotestosterone on oxidative stress and carotenoid-dependent colour expression in a bird. *Behav. Ecol. Sociobiol.*, 66, 1319–31.

Catry, I., Dias, M., Catry, T., et al. (2011). Individual variation in migratory movements and winter behaviour of Iberian lesser kestrels *Falco naumanni* revealed by geolocators. *Ibis*, 153, 154–64.

Catry, I., Amano, T., Franco, A. M. A. and Sutherland, W. J. (2012). Influence of spatial and temporal dynamics of agricultural practices on the globally endangered lesser kestrel. *J. Appl. Ecol.*, 144, 1111–9.

Catry, T., Moreira, F., Alcazar, R., Rocha, P. A. and Catry, I. (2017). Mechanisms and fitness consequences of laying decisions in a migratory raptor. *Behav. Ecol.*, 28, 222–32.

Cavé, A. J. (1967). The breeding of the kestrel in the reclaimed area Oostelijk Flevoland. *Nether. J. Zool.*, 18, 313–407.

Cenizo, M., Noriega, J. I. and Reguero, M. A. (2016). A stem falconid bird from the Lower Eocene of Antarctica and the early southern radiation of the falcons. *J. Ornithol.*, 157, 885–94.

Cerling, T. E., Harris, J. M., MacFadden, B. J., et al. (1997). Global vegetation change through the Miocene/Pliocene boundary. *Nature*, 389, 153–8.

Charter, M., Izhaki, I., Bouskila, A. and Leshem, Y. (2007a). Breeding success of the Eurasian kestrel (*Falco tinnunculus*) nesting on buildings in Israel. *J. Rapt. Res.*, 41, 139–43.

Charter, M., Izhaki, I., Bouskila, A. and Leshem, Y. (2007b). The effect of different nest types on the breeding success of the Eurasian Kestrel (*Falco tinnunculus*) in a rural ecosystem. *J. Rapt. Res.*, 41, 143–9.

Charter, M., Izhaki, I. and Leshem, Y. (2010). Effects of risk of competition and predation on large secondary cavity breeders. *J. Ornithol.*, 151, 791–5.

Christians, J. K. (2002). Avian egg size: variation within species and inflexibility within individuals. *Biol. Rev.*, 77, 1–26.

Clark, A. B. and Wilson, D. S. (1981). Avian breeding adaptations: hatching asynchrony, brood reduction and nest failure. *Q. Rev. Biol.*, 56, 253–77.

Clegg, T. M. (1971). Kestrel hiding prey. *Scot. Bird*, 6, 276–7.

Clegg, T. M. and Henderson, D. S. (1971). Kestrel taking prey from short-eared owl. *Brit. Birds*, 64, 317–8.

Collar, N. J. and Stuart, S. N. (1985). *Threatened birds of Africa and related islands: the ICBP/ IUCN Red Data book*. Cambridge, UK: International Council for Bird Preservation, and International Union for Conservation of Nature and Natural Resources.

Collopy, M. W. (1977). Food caching by female American kestrels in winter. *Condor*, 79, 63–8.

Cooke, A. S., Bell, A. A. and Haas, M. B. (1982). *Predatory birds, pesticides and pollution*. Cambridge, UK: Institute of Terrestrial Ecology.

Costantini, D. (2014). *Oxidative stress and hormesis in evolutionary ecology and physiology: a marriage between mechanistic and evolutionary approaches* (p. 348). Berlin: Springer-Verlag.

Costantini, D. and Dell'Omo, G. (2006a). Effects of T-cell-mediated immune response on avian oxidative stress. *Comp. Biochem. Physiol. Part A*, 145, 137–42.

Costantini, D. and Dell'Omo, G. (2006b). Environmental and genetic components of oxidative stress in wild kestrel nestlings (*Falco tinnunculus*). *J. Comp. Physiol. B*, 176, 575–9.

Costantini, D. and Dell'Omo, G. (2010). Sex-specific predation on two lizard species by kestrels. *Russ. J. Ecol.*, 41, 99–101.

Costantini, D., Casagrande, S., Di Lieto, G., Fanfani, A. and Dell'Omo, G. (2005). Consistent differences in feeding habits between neighbouring breeding kestrels. *Behaviour*, 142, 1409–21.

Costantini, D., Casagrande, S., De Filippis, S., et al. (2006). Correlates of oxidative stress in wild kestrel nestlings (*Falco tinnunculus*). *J. Comp. Physiol. B*, 176, 329–37.

Costantini, D., Casagrande, S. and Dell'Omo, G. (2007a). MF magnitude does not affect body condition, pro-oxidants and anti-oxidants in Eurasian kestrel (*Falco tinnunculus*) nestlings. *Environm. Res.*, 104, 361–6.

Costantini, D., Fanfani, A. and Dell'Omo, G. (2007b). Carotenoid availability does not limit the capability of nestling kestrels (*Falco tinnunculus*) to cope with oxidative stress. *J. Exp. Biol.*, 210, 1238–44.

Costantini, D., Bruner, E., Fanfani, A. and Dell'Omo, G. (2007c). Male-biased predation of western green lizards by Eurasian kestrels. *Naturwissenschaften*, 94, 1015–20.

Costantini, D., Coluzza, C., Fanfani, A. and Dell'Omo, G. (2007d). Effects of carotenoid supplementation on colour expression, oxidative stress and body mass in rehabilitated captive adult kestrels (*Falco tinnunculus*). *J. Comp. Physiol. B*, 177, 723–31.

Costantini, D., Fanfani, A. and Dell'Omo, G. (2008). Effects of corticosteroids on oxidative damage and circulating carotenoids in captive adult kestrels (*Falco tinnunculus*). *J. Comp. Physiol. B*, 178, 829–35.

Costantini, D., Bertacche, V., Pastura, B. and Turk, A. (2009a). Dehydrolutein: a metabolically derived carotenoid never observed in raptors. *Curr. Zool.*, 55, 238–42.

Costantini, D., Casagrande, S., Carello, L. and Dell'Omo, G. (2009b). Body condition variation in kestrel (*Falco tinnunculus*) nestlings in relation to breeding conditions. *Ecol. Res.*, 24, 1213–21.

Costantini, D., Carello, L. and Dell'Omo, G. (2010a). Patterns of covariation among weather conditions, winter North Atlantic Oscillation index, and reproductive traits in Mediterranean kestrels (*Falco tinnunculus*). *J. Zool.*, 280, 177–84.

Costantini, D., Carello, L. and Dell'Omo, G. (2010b). Temporal covariation of egg volume and breeding conditions in the common kestrel (*Falco tinnunculus*) in the Mediterranean region. *Ornis Fenn.*, 87, 144–52.

Costantini, D., Dell'Omo, G., La Fata, I. and Casagrande, S. (2014). Reproductive performance of Eurasian Kestrel Falco tinnunculus in an agricultural landscape with a mosaic of land uses. *Ibis*, 156, 768–76.

Costantini, D., Casasole, G., AbdElgawad, H., Asard, H. and Eens, M. (2016). Experimental evidence that oxidative stress influences reproductive decisions. *Funct. Ecol.*, 30, 1169–74.

Costantini, D., Gustin, M., Ferrarini, A. and Dell'Omo, G. (2017a). Estimates of avian collision with power lines and carcass disappearance across differing environments. *Anim. Cons.*, 20, 173–81.

Costantini, D., Sebastiano, M., Goossens, B. and Stark, D. (2017b). Jumping in the night: an investigation of leaping activity of western tarsier (*Cephalopachus bancanus borneanus*) using accelerometers. *Folia Primatol.*, 88, 46–56.

Costantini, D., Blévin, P., Herzke, D., et al. (2019). Higher plasma oxidative damage and lower plasma antioxidant defences in an Arctic seabird exposed to longer perfluoroalkyl acids. *Environ. Res.*, 168, 278–85.

Cramp, S. and Simmons, K. E. L. (1980). *Handbook of the birds of Europe, the Middle East and North Africa. Vol. II, Hawks to bustards*. Oxford, UK: Oxford University Press.

Csermely, D., Casagrande, S. and Calimero, A. (2006). Differential defensive response of common kestrels against a known or unknown predator. *Ital. J. Zool.*, 73, 125–8.

Culina, A., Radersma, R. and Sheldon, B. C. (2015). Trading up: the fitness consequences of divorce in monogamous birds. *Biol. Rev.*, 90, 1015–34.

Cunningham, E. J. A. and Birkhead, T. R. (1998). Sex roles and sexual selection. *Anim. Behav.*, 56, 1311–21.

Daan, S. and Dijkstra, C. (1988). Date of birth and reproductive value of kestrel eggs: on the significance of early breeding (pp. 85–114). In C. Dijkstra, ed., *Reproductive tactics in the kestrel Falco tinnunculus: a study in evolutionary biology*. PhD thesis, University of Groningen, Groningen, the Netherlands.

Daan, S., Dijkstra, C., Drent, R. and Meijer, T. (1989a). Food supply and the annual timing of avian reproduction. In H. Ouellet, ed., *Acta XIX Congressus Internationalis Ornithologici*. Volume I, Proceedings XIX International Ornithological Congress (pp. 392–407). Ottawa: University of Ottawa Press.

Daan, S., Masman, D., Strijkstra, A. and Verhulst, S. (1989b). Intraspecific allometry of basal metabolic rate: relations with body size, temperature, composition and circadian phase in the kestrel. *J. Biol. Rhythms*, 4, 267–83.

Daan, S., Dijkstra, C. and Tinbergen, J. M. (1990). Family planning in the kestrel: the ultimate control of covariation of laying date and clutch size. *Behaviour*, 114, 83–116.

Daan, S., Deerenberg, C. and Dijkstra, C. (1996). Increased daily work precipitates natural death in the kestrel. *J. Anim. Ecol.*, 65, 539–44.

Danchin, E., Charmantier, A., Champagne, F. A., et al. (2011). Beyond DNA: integrating inclusive inheritance into an extended theory of evolution. *Nat. Rev. Genet.*, 12, 475–86.

Darwin, C. (1871). *The descent of man, and selection in relation to sex*. London: John Murray.

Davies, T. (1975). Food of the kestrel in the winter and early spring. *Bird Study*, 22, 85–92.

Dawkins, R. (1982). *The extended phenotype*. Oxford, UK: Oxford University Press.

Dawson, A. and Goldsmith, A. (1982). Prolactin and gona-dotrophin secretion in wild starlings (*Sturnus vulgaris*) during the annual cycle and in relation to nesting, incubation and rearing young. *Gen. Comp. Endocrinol.*, 48, 213–21.

Dawson, A. and Goldsmith, A. (1985). Modulation of gonadotrophin and prolactin secretion by daylength and breeding behaviour in free-lining starlings, Sturnus vulgaris. *J. Zool.*, 206, 241–52.

Dawson, R. D. and Bortolotti, G. R. (1997). Ecology of parasitism of nestling American kestrels by *Carnus hemapterus* (Diptera: Carnidae). *Can. J. Zool.*, 75, 2021–6.

Dawson, R. D. and Bortolotti, G. R. (2000). Reproductive success of American kestrels: the role of prey abundance and weather. *Condor*, 102, 814–22.

Dawson, R. D. and Bortolotti, G. R. (2002). Experimental evidence for food limitation and sex-specific strategies of American kestrels (*Falco sparverius*) provisioning offspring. *Behav. Ecol. Sociobiol.*, 52, 43–52.

Dawson, R. D. and Bortolotti, G. R. (2003). Parental effort of American kestrels: the role of variation in brood size. *Can. J. Zool.*, 81, 852–60.

Dawson, R. D. and Bortolotti, G. R. (2008). Experimentally prolonging the brood-rearing period reveals sex-specific parental investment strategies in American kestrels (*Falco sparverius*). *The Auk*, 125, 889–95.

De Neve, L., Fargallo, J. A., Vergara, P., et al. (2008). Effects of maternal carotenoid availability in relation to sex, parasite infection and health status of nestling kestrels (*Falco tinnunculus*). *J. Exp. Biol.*, 211, 1414–25.

Deerenberg, C., Pen, I., Dijkstra, C., et al. (1995). Parental energy expenditure in relation to manipulated brood size in the European kestrel, *Falco tinnunculus*. *Zoology*, 99, 38–47.

Delibes, M., Gaona, P. and Ferreras, P. (2001). Effects of an attractive sink leading into maladaptive habitat selection. *Am. Nat.*, 158, 277–85.

Dell'Omo, G., Costantini, D., Di Lieto, G. and Casagrande, S. (2005). Gli uccelli e le linee elettriche. *Alula*, 12, 103–14.

Dell'Omo, G., Costantini, D., Wright, J., Casagrande, S. and Shore, R. F. (2008). PCBs in the eggs of Eurasian kestrels indicate exposure to local pollution. *Ambio*, 37, 452–6.

Dell'Omo, G., Costantini, D., Lucini, V., et al. (2009). Magnetic fields produced by power lines do not affect growth, serum melatonin, leukocytes and fledging success in wild kestrels. *Comp. Biochem. Physiol. Part C*, 150, 372–6.

deMent, S. H., Rikard, S. T. and Wommack, E. A. (2014). Genetic evaluation for *Falco sparverius paulus* within breeding American Kestrels in the Midlands/Sandhills region of South Carolina. *The Oriole*, 79, 1–16.

DeNiro, M. J. and Epstein, S. (1978). Influence of diet on the distribution of carbon isotopes in animals. *Geochim. Cosmochim. Acta*, 42, 495–506.

DeNiro, M. J. and Epstein, S. (1981). Influence of diet on the distribution of nitrogen isotopes in animals. *Geochim. Cosmochim. Acta*, 45, 341–51.

Deschamps, P. (1980). Point local sur l'électrocution des rapaces (et autres oiseaux) sur les lignes à moyenne tension de la région grenobloise. *La Niverolle*, 5, 59–66.

Dewar, S. M. and Shawyer, C. R. (1996). *Boxes, baskets and platforms: artificial nest sites for owls and other birds of prey*. London: Chelmsford Press.

Dhondt, A., Adriaensen, F. and Verwimp, N. (1997). Are Belgian kestrels *Falco tinnunculus* migratory: an analysis of ringing recoveries. *Ring. Migrat.*, 18, 91–101.

Dickinson, E. C. (2003). *The Howard and Moore complete checklist of the birds of the world*, 3rd edn. Princeton, NJ: Princeton University Press.

Dietz, M. W., Daan, S. and Masman, D. (1992). Energy requirements for molt in the kestrel *Falco tinnunculus*. *Physiol. Zool.*, 65, 1217–35.

Dijkstra, C. (1988). *Reproductive tactics in the kestrel, Falco tinnunculus*. PhD thesis, University of Groningen, Groningen, the Netherlands.

Dijkstra, C., Vuursteen, L., Daan, S. and Masman, D. (1982). Clutch size and laying date in the kestrel *Falco tinnunculus*: effect of food supplementary food. *Ibis*, 124, 211–3.

Dijkstra, C., Daan, S. and Buker, J. B. (1990a). Adaptive seasonal variation in the sex ratio of kestrel broods. *Funct. Ecol.*, 4, 143–7.

Dijkstra, C., Bult, A., Bijlsma, S., Daan, S., Meijer, T. and Zijlstra, M. (1990b). Brood size manipulations in the kestrel *Falco tinnunculus*: effects on offspring and adult survival. *J. Anim. Ecol.*, 59, 269–85.

Dixon, A., Rahman, Lutfor MD, Galtbalt, B., et al. (2019). Mitigation techniques to reduce avian electrocution rates. *Wildlife Soc. Bull.*, 43, 476–83.

Dmitriew, C. M. (2011). The evolution of growth trajectories: what limits growth rate? *Biol. Rev.*, 86, 97–116.

Dogliero, A., Rota, A., Lofiego, R., Mauthe von Degerfeld, M. and Quaranta, G. (2016). Semen evaluation in four autochthonous wild raptor species using computer-aided sperm analyzer. *Theriogenol.*, 85, 1113–7.

Drent, E. H. and Daan, S. (1980). The prudent parent: energetic adjustments in avian breeding. *Ardea*, 68, 225–52.

du Feu, C. R., Joys, A. C., Clark, J. A., et al. (2009). *EURING Data Bank geographical index 2009* (www.euring.org/edb).

Duke, G. E., Evanson, O. A. and Jegers, A. (1976). Meal to pellet intervals in 14 species of captive raptors. *Comp. Biochem. Physiol. Part A*, 53, 1–6.

Duke, G. E., Tererick, A. L., Reynhout, J. K., Bird, D. M. and Place, A. E. (1996). Variability among individual American kestrels (*Falco sparverius*) in parts of day-old chicks eaten, pellet size, and pellet egestion frequency. *J. Raptor Res.*, 30, 213–8.

Duncan, J. R. and Bird, D. M. (1989). The influence of relatedness and display effort on the mate choice of captive female American kestrels. *Anim. Behav.*, 37, 112–7.

Durany, E., Garcia, S. and Santaeufemia, X. (2003). Los cernìcalos urbanos de Barcelona. *Quercus*, 204, 24–7.

Eagles-Smith, C. A., Silbergeld, E. K., Basu, N., et al. (2018). Modulators of mercury risk to wildlife and humans in the context of rapid global change. *Ambio*, 47, 170–97.

Edwards, N. P., van Veelen, A., Anné, J., et al. (2016). Elemental characterisation of melanin in feathers via synchrotron X-ray imaging and absorption spectroscopy. *Sci. Rep.*, 6, 34002.

Eens, M., Van Duyse, E. V., Berghman, L. and Pinxten, R. (2000). Shield characteristics are testosterone-dependent in both male and female moorhens. *Horm. Behav.*, 37, 126–34.

Eeva, T., Sillanpää, S. and Salminen, J.-P. (2009). The effects of diet quality and quantity on plumage colour and growth of great tit *Parus major* nestlings: a food manipulation experiment along a pollution gradient. *J. Avian Biol.*, 40, 491–9.

Ehleringer, J. R. and Rundel, P. W. (1988). Stable isotopes: history, units, and instrumentation. In P. W. Rundel, J. R. Ehleringer and K. A. Nagy, eds., *Stable isotopes in ecological research (Ecological Studies 68)* (pp. 1–16). Berlin: Springer.

El Agamey, A., Lowe, G. M., McGarvey, D. J., et al. (2004). Carotenoid radical chemistry and antioxidant/pro-oxidant properties. *Arch. Biochem. Biophys.*, 430, 37–48.

El Halawani, M. E., Silsby, J. L., Behnke, E. J. and Fehrer, S. C. (1986). Hormonal induction of incubation behavior in ovariectomized female turkeys (*Meleagris gallopavo*). *Biol. Reprod.*, 35, 59–67.

El Halawani, M. E., Silsby, J. L., Youngren, O. M. and Phillips, R. E. (1991). Exogenous prolactin delays photo-induced sexual maturity and suppresses ovariectomy-induced luteinizing hormone secretion in the turkey (*Meleagris gallopavo*). *Biol. Reprod.*, 44, 420–4.

Elliott, D. (1971). Kestrel apparently robbing weasel of a vole. *Brit. Birds*, 64, 229.

Ericson, P. G. P. (2012). Evolution of terrestrial birds in three continents: biogeography and parallel radiations. *J. Biogeogr.*, 39, 813–24.

Ericson, P. G. P., Anderson, C. L., Britton, T., et al. (2006). Diversification of Neoaves: integration of molecular sequence data and fossils. *Biol. Lett.*, 2, 543–7.

Eriksson, U., Roos, A., Lind, Y., et al. (2016). Comparison of PFASs contamination in the freshwater and terrestrial environments by analysis of eggs from osprey (*Pandion haliaetus*), tawny owl (*Strix aluco*), and common kestrel (*Falco tinnunculus*). *Environ. Res.*, 149, 40–7.

Erlinge, S., Göransson, G., Hansson, L., et al. (1983). Predation as a regulating factor on small rodent populations in southern Sweden. *Oikos*, 40, 36–52.

Everett, M. J. (1968). Kestrel taking prey from barn owl. *Brit. Birds*, 61, 264.

Falconer, D. S. (1981). *Introduction to quantitative genetics*. London: Longman.

Fallacara, D. M., Halbrook, R. S. and French, J. B. (2011). Toxic effects of dietary methylmercury on immune function and hematology in American kestrels (*Falco sparverius*). *Environ. Toxicol. Chem.*, 30, 1320–7.

Fargallo, J. A., Blanco, G., Potti, J. and Viñuela, J. (2001). Nestbox provisioning in a rural population of Eurasian kestrels: breeding performance, nest predation and parasitism. *Bird Study*, 48, 236–44.

Fargallo, J. A., Laaksonen, T., Pöyri, V. and Korpimäki, E. (2002). Inter-sexual differences in the immune response of Eurasian kestrel nestlings under food shortage. *Ecol. Lett.*, 5, 95–101.

Fargallo, J. A., Laaksonen, T., Korpimäki, E., et al. (2003). Size-mediated dominance and begging behaviour in Eurasian kestrel broods. *Evol. Ecol. Res.*, 5, 549–58.

Fargallo, J. A., Laaksonen, T., Korpimäki, E. and Wakamatsu, K. (2007a). A melanin-based trait reflects environmental growth conditions of nestling male Eurasian kestrels. *Evol. Ecol.*, 21, 157–71.

Fargallo, J. A., Martínez-Padilla, J., Toledano-Díaz, A., Santiago-Moreno, J. and Dávila, J. A. (2007b). Sex and testosterone effects on growth, immunity and melanin coloration of nestling Eurasian kestrels. *J. Anim. Ecol.*, 76, 201–9.

Fargallo, J. A., Martínez-Padilla, J., Viñuela, J., et al. (2009). Kestrel–prey dynamic in a Mediterranean region: the effect of generalist predation and climatic factors. *PLoS ONE*, 4, e4311.

Fargallo, J. A., López-Rull, I., Mikšík, I., Eckhardt, A. and Peralta-Sánchez, J. M. (2014). Eggshell pigmentation has no evident effects on offspring viability in common kestrels. *Evol. Ecol.*, 28, 627–37.

Farmer, G. C., McCarty, K., Robertson, S., Robertson, B. and Bildstein, K. L. (2006). Suspected predation by accipiters on radio-tracked American kestrels (*Falco sparverius*) in eastern Pennsylvania, U.S.A. *J. Rapt. Res.*, 40, 294–7.

Fattorini, S., Manganaro, A., Piattella, E. and Salvati, L. (1999). Role of the beetles in raptor diets from a Mediterranean urban area. *Fragm. Entomol.*, 31, 57–69.

Fay, R., Michler, S., Laesser, J. and Schaub, M. (2019). Integrated population model reveals that kestrels breeding in nest boxes operate as a source population. *Ecography*, 42, 2122–31.

Feare, C. J., Temple, S. A. and Procter, J. (1974). The status, distribution and diet of the Seychelles kestrel (*Falco araea*). *Ibis*, 116, 548–51.

Fennel, C. M. (1954). Notes on the nesting of the kestrel in Japan. *Condor*, 56, 106–7.

Fernie, K. J. and Bird, D. M. (2000). Effects of electromagnetic fields on the growth of nestling American kestrels. *Condor*, 102, 461–5.

Fernie, K. J. and Bird, D. M. (2001). Evidence of oxidative stress in American kestrels exposed to electromagnetic fields. *Environ. Res.*, 86, 198–207.

Fernie, K. J. and Marteinson, S. C. (2016). Sex-specific changes in thyroid gland function and circulating thyroid hormones in nestling American kestrels (*Falco sparverius*) following embryonic exposure to polybrominated diphenyl ethers by maternal transfer. *Environ. Toxicol. Chem.*, 35, 2084–91.

Fernie, K. J., Bird, D. M. and Petitclerc, D. (1999). Effects of electromagnetic fields on photophasic circulating melatonin levels in American kestrels. *Environ. Health Perspect.*, 107, 901–4.

Fernie, K. J., Bird, D. M., Dawson, R. D. and Laguë, P. C. (2000). Effects of electromagnetic fields on the reproductive success of American kestrels. *Physiol. Biochem. Zool.*, 73, 60–5.

Fernie, K. J., Smits, J. E., Bortolotti, G. R. and Bird, D. M. (2001). Reproduction success of American Kestrels exposed to dietary polychlorinated biphenyls. *Environ. Toxicol. Chem.*, 20, 776–81.

Fernie, K. J., Mayne, G., Shutt, J. L., et al. (2005). Evidence of immunomodulation in nestling American kestrels (*Falco sparverius*) exposed to environmentally relevant PBDEs. *Environ. Pollut.*, 138, 485–93.

Ferrer, M. (2012). *Birds and power lines: from conflict to solution*. Seville: Endesa S. A. and Fundación Migres.

Figueroa, R. and Corales, E. (2002). Winter diet of the American kestrel (*Falco sparverius*) in the forested Chilean Patagonia, and its relation to the availability of prey. *Inter. Hawk-watcher*, 5, 7–14.

Folstad, I. and Karter, A. J. (1992). Parasites bright males and the immunocompetence handicap. *Am. Nat.*, 139, 603–22.

Forbes, N. A. and Fox, M. T. (2005). Field trial of a *Caryospora* species vaccine for controlling clinical coccidiosis in falcons. *Veter. Rec.*, 156, 134–8.

Forbes, N. A. and Simpson, G. N. (1997). *Caryospora neofalconis*: an emerging threat to captive bred raptors in the United Kingdom. *J. Avian Med. Surg.*, 11, 110–4.

Fraissinet, M. (2008). La frequentazione urbana delle specie del genere *Falco* in Italia e in Europa. Una monografia. *Ecol. Urb.*, 20, 29–56.

Fransson, T., Jansson, L., Kolehmainen, T., Kroon, C. and Wenninger, T. (2017) *EURING list of longevity records for European birds*. Available from https://euring.org/data-and-codes/long evity-list.

Fritz, H. (1998). Wind speed as a determinant of kleptoparasitism by Eurasian Kestrel *Falco tinnunculus* on Short-eared Owl *Asio flammeus*. *J. Avian Biol.*, 29, 331–3.

Fry, B. and Sherr, E. B. (1988). δI3C measurements as indicators of carbon flow in marine and freshwater ecosystems. In P. W. Rundel, J. R. Ehleringer and K. A. Nagy, eds., *Stable isotopes in ecological research* (pp. 196–229). New York: Springer-Verlag.

Fuchs, J., Johnson, J. A. and Mindell, D. P. (2012). Molecular systematics of the caracaras and allies (Falconidae: Polyborinae) inferred from mitochondrial and nuclear sequence data. *Ibis*, 154, 520–32.

Fuchs, J., Johnson, J. A. and Mindell, D. P. (2015). Rapid diversification of falcons (Aves: Falconidae) due to expansion of open habitats in the Late Miocene. *Molec. Phylogen. Evol.*, 82, 166–82.

García-Borrón, J. C. and Olivares Sánchez, M. C. (2011). Biosynthesis of melanins. In J. Borovanský and P. A. Riley, eds., *Melanins and melanosomes: biosynthesis, biogenesis, physiological, and pathological functions* (pp. 87–116). Weinheim: Wiley-Blackwell.

Gard, N. W. and Bird, D. M. (1990). Breeding behavior of American kestrels raising manipulated brood sizes in years of varying prey abundance. *Wilson Bull.*, 102, 605–14.

Garratt, C. M., Hughes, M., Eagle, G., et al. (2011). Foraging habitat selection by breeding common kestrels *Falco tinnunculus* on lowland farmland in England. *Bird Study*, 58, 90–8.

Gaymer, R. (1967). Observations on the birds of Aldabra in 1964 and 1965. *Atoll Res. Bull.*, 118, 113–25.

Gaymer, R., Blackman, R. A. A., Dawson, P. G., Penny, M. J. and Penny, C. M. (1969). The endemic birds of Seychelles. *Ibis*, 111, 157–76.

Genelly, R. E. (1978). Observations of the Australian kestrel on northern Tablelands of New South Wales. *Emu*, 78, 137–44.

Geng, R., Zhang, X., Ou, W., et al. (2009). Diet and prey consumption of breeding common kestrel (*Falco tinnunculus*) in Northeast China. *Prog. Nat. Sci.*, 19, 1501–7.

Géroudet, P. (1978). *Les rapaces diurnes et nocturnes d'Europe*. Paris: Delachaux and Niestlé.

Gil-Delgado, J. A., Verdejo, J. and Barba, E. (1995). Nestling diet and fledgling production of Eurasian kestrels (*Falco tinnunculus*) in eastern Spain. *J. Raptor Res.*, 29, 240–4.

Gill, F. and Donsker, D. (2018). IOC World Bird List (v. 8.2). Doi:10.14344/IOC.ML.8.2.

Giraldeau, L.-A. and Lefebvre, L. (1985). Individual feeding preferences in feral groups of rock doves. *Can. J. Zool.*, 63, 189–91.

Gomes, A. C. R. and Cardoso, G. C. (2018). Choice of high-quality mates versus avoidance of low-quality mates. *Evolution*, 72, 608–16.

Gómez-Ramírez, P., Shore, R. F., van den Brink, N. W., et al. (2014). An overview of existing raptor contaminant monitoring activities in Europe. *Environ. Int.*, 67, 12–21.

Goodland, R. (1973). Ecological perspectives of power transmission. In R. Goodland, ed., *Power lines and the environment* (pp. 1–35). Millbrook, NY: The Cary Arboretum of the New York Botanical Garden.

Gosler, A. G., Connor, O. R. and Bonser, R. H. C. (2011). Protoporphyrin and eggshell strength: preliminary findings from a passerine bird. *Avian Biol. Res.*, 4, 214–23.

Graf, P. M., Wilson, R. P., Qasem, L., Hackländer, K. and Rosell, F. (2015). The use of acceleration to code for animal behaviours; a case study in free-ranging Eurasian beavers *Castor fiber. PLoS ONE*, 10, e0136751.

Graham, D. L., Maré, C. J., Ward, F. P. and Peckham, M. C. (1975). Inclusion body disease (herpesvirus infection) of falcons (IBDF). *J. Wildl. Dis.*, 11, 83–91.

Grant, C. H. B. and Mackworth-Praed, C. W. (1934). On the races and distribution of the African and Arabian Kestrels of the *Falco tinnunculus* group, with descriptions of two new races. *Bull. Brit. Ornith. Club*, 54, 75–83.

Grant, P. R. (1982). Variation in the size and shape of Darwin's finch eggs. *Auk*, 99, 15–23.

Greenwood, A. G. and Cooper, J. E. (1982). Herpesvirus infections in falcons. *Veter. Rec.*, 111, 514.

Greenwood, P. J. and Harvey, P. H. (1982). The natal and breeding dispersal of birds. *Annu. Rev. Ecol. System.*, 13, 1–21.

Griffiths, C. S. (1999). Phylogeny of the *Falconidae* inferred from molecular and morphological data. *The Auk*, 116, 116–30.

Griffiths, C. S., Barrowclough, G. F., Groth, J. G. and Mertz, L. (2004). Phylogeny of the Falconidae (Aves): a comparison of the efficacy of morphological, mitochondrial, and nuclear data. *Molec. Phylogen. Evol.*, 32, 101–9.

Griggio, M., Hamerstrom, F., Rosenfield, R. N. and Tavecchia, G. (2002). Seasonal variation in sex ratio of fledgling American Kestrels: a long term study. *Wilson Bull.*, 114, 474–8.

Groombridge, J. J, Jones, C. G., Bruford, M. W. and Nichols, R. A. (2000). 'Ghost' alleles of the Mauritius kestrel. *Nature*, 403, 616.

Groombridge, J. J., Jones, C. G., Bayes, M. K., et al. (2002). A molecular phylogeny of African kestrels with reference to divergence across the Indian Ocean. *Molec. Phylogen. Evol.*, 25, 267–77.

Groombridge, J. J., Dawson, D. A., Burke, T., et al. (2009). Evaluating the demographic history of the Seychelles Kestrel *Falco araea*: genetic evidence for recovery from a population bottleneck following minimal conservation management. *Biol. Conserv.*, 142, 2250–7.

Grünewälder, S., Broekhuis, F., Macdonald, D. W., et al. (2012). Movement activity based classification of animal behaviour with an application to data from cheetah (*Acinonyx jubatus*). *PLoS ONE*, 7, e49120.

Guigueno, M. F., Karouna-Renier, N. K., Henry, P. F. P., et al. (2018). Sex-specific responses in neuroanatomy of hatchling American kestrels in response to embryonic exposure to the flame retardants bis(2-ethylhexyl)-2,3,4,5-tetrabromophthalate and 2-ethylhexyl-2,3,4,5-tetrabro-mobenzoate. *Environ. Toxicol. Chem.*, 37, 3032–40.

Hachisuka, M. (1953). *The dodo and kindred birds or the extinct birds of the Mascarene Islands.* London: HF and G Witherby.

Hagen, I. (1969). Norske undersøkelser over avkomproduksjonen hos rovfugler og ugler sett i relasjon til smågnagerbestandens vekslinger. *Fauna*, 22, 73–126.

Hagen, Y. (1952). *Rovfuglene og viltpleien.* Oslo: Gyldendal Norsk Forlag.

Hakkarainen, H. and Korpimäki, E. (1996). Competitive and predatory interactions among raptors: an observational and experimental study. *Ecology*, 77, 1134–42.

Hakkarainen, H., Korpimäki, E., Huhta, E. and Palokangas, P. (1993). Delayed maturation in plumage colour: evidence for the female-mimicry hypothesis in the kestrel. *Behav. Ecol. Sociobiol.*, 33, 247–51.

Hakkarainen, H., Huhta, E., Lahti, K., et al. (1996). A test of male mating and hunting success in the kestrel: the advantages of smallness? *Behav. Ecol. Sociobiol.*, 39, 375–80.

Hall, J. S., Ip, H. S., Franson, J. C., et al. (2009). Experimental infection of a North American raptor, American kestrel (*Falco sparverius*), with highly pathogenic avian influenza virus (H5N1). *PLoS ONE*, 4, e7555.

Hall, M. R. and Goldsmith, A. R. (1983). Factors affecting prolactin secretion during breeding and incubation in the domestic duck (*Anas platyrhynchos*). *Gen. Comp. Endocrinol.*, 49, 270–6.

Hamilton, W. D. and Zuk, M. (1982). Heritable true fitness and bright birds: a role for parasites? *Science*, 218, 384–7.

Hammond, T. T., Springthorpe, D., Walsh, R. E. and Berg-Kirkpatrick, T. (2016). Using accelerometers to remotely and automatically characterize behavior in small animals. *J. Exp. Biol.*, 219, 1618–24.

Hanski, I., Henttonen, H., Korpimäki, E., Oksanen, L. and Turchin, P. (2001). Small rodent dynamics and predation. *Ecology*, 82, 1505–20.

Hart, J. (1972). Food habits of American kestrels in a low vole year. *Raptor Res.*, 6, 1–3.

Hartert, E. (1912–1921). *Die Vögel der paläarktischen Fauna. Systematische Übersicht der in Europa, Nord-Asien und der Mittelmeerregion vorkommenden Vögel*. Vol. II. Berlin: Verlag von R. Friedländer und Sohn.

Hartley, R. C. and Kennedy, M. W. (2004). Are carotenoids a red herring in sexual display? *Trends Ecol. Evol.*, 19, 353–4.

Hasenclever, H., Kostrzewa, A. and Kostrzewa, R. (1989). Brutbiologie des turmfalken (*Falco tinnunculus*): 16 jährige untersuchungen in Westfalen. *J. Ornith.*, 130, 229–37.

Hearing, V. J. (1993). Unraveling the melanocyte. *Am. J. Hum. Genet.*, 52, 1–7.

Heath, J. (1997). Corticosterone levels during nest departure of juvenile American kestrels. *The Condor*, 99, 806–11.

Heim de Balsac, H. and Mayaud, N. (1962). *Les oiseaux du Nord-Ouest de l'Afrique*. Paris: Paul Lechevalier.

Hendry, A. P. and Day, T. (2005). Population structure attributable to reproductive time: isolation by time and adaptation by time. *Mol. Ecol.*, 14, 901–16.

Hernández-Lambraño, R. E., Sánchez-Agudo, J. A. and Carbonell, R. (2018). Where to start? Development of a spatial tool to prioritise retrofitting of power line poles that are dangerous to raptors. *J. Appl. Ecol.*, 55, 2685–97.

Hernández-Matías, A., Real, J., Moleón, M., et al. (2013). From local monitoring to a broad-scale viability assessment: a case study for the Bonelli's Eagle in western Europe. *Ecol. Monogr.*, 83, 239–61.

Hernández-Pliego, J., Rodríguez, C., Dell'Omo, G. and Bustamante, J. (2017). Combined use of tri-axial accelerometers and GPS reveals the flexible foraging strategy of a bird in relation to weather conditions. *PLoS ONE*, 12, e0177892.

Hickey, J. J. and Anderson, D. W. (1968). Chlorinated hydrocarbons and eggshell changes in raptorial and fish-eating birds. *Science*, 162, 271–3.

Hill, G. E. and McGraw, K. J. (2006a). *Bird coloration. Volume I. Mechanisms and measurements*. Cambridge, MA: Harvard University Press.

Hill, G. E. and McGraw, K. J. (2006b). *Bird coloration. Volume II. Function and evolution*. Cambridge, MA: Harvard University Press.

Hille, S. M. (2002). *Sexual dimorphism and niche differentiation in island populations of the kestrel.* PhD thesis, University of Vienna, Vienna, Austria.

Hille, S. M., Nesje, M. and Segelbacher, G. (2003). Genetic structure of kestrel populations and colonization of the Cape Verde archipelago. *Mol. Ecol.*, 12, 2145–51.

Hille, S. M., Nash, J. P. and Krone, O. (2007). Hematozoa in endemic subspecies of common kestrel in the Cape Verde Islands. *J. Wildl. Dis.*, 43, 752–7.

Hoffman, D. J., Melancon, M. J., Klein, P. N., et al. (1996). Developmental toxicity of PCB 126 (3,3′,4,4′,5-pentachlorobiphenyl) in nestling American kestrels (*Falco sparverius*). *Fundam. Appl. Toxicol.*, 34, 188–200.

Hoffman, D. J., Melancon, M. J., Klein, P. N., Eisemann, J. D. and Spann, J. W. (1998). Comparative and developmental toxicity of planar polychlorinated biphenyl congeners in chickens, American kestrels, and common terns. *Environ. Toxicol. Chem.*, 17, 747–57.

Holte, D., Köppen, U., and Schmitz-Ornés, A. (2016). Partial migration in a central European raptor species: an analysis of ring re-encounter data of common kestrels *Falco tinnunculus*. *Acta Ornithol.*, 51, 39–54.

Hoogenboom, I. and Dijkstra, C. (1987). *Sarcocystis cernae*: a parasite increasing the risk of predation of its intermediate host, Microtus arvalis. *Oecologia*, 74, 86–92.

Hunt, K. and Wingfield, J. (2004). Effect of estradiol implants on reproductive behavior of female Lapland longspurs (*Calcarius lapponicus*). *Gen. Comp. Endocrinol.*, 137, 248–62.

Hustler, K. (1983). Breeding biology of the Greater Kestrel. *Ostrich*, 54, 129–40.

Hyuga, I. (1956). Breeding colonies of Japanese kestrels. *Tori*, 14, 17–24.

Ille, R., Hoi, H., Grinschgl, F. and Zink, R. (2002). Paternity assurance in two species of colonially breeding falcon: the kestrel *Falco tinnunculus* and the red-footed falcon *Falco vespertinus*. *Etologia*, 10, 11–5.

Inger, R. and Bearhop, S. (2008). Applications of stable isotope analyses to avian ecology. *Ibis*, 150, 447–61.

Janss, G. F. E. (2000). Avian mortality from power lines: a morphologic approach of a species-specific mortality. *Biol. Cons.*, 95, 353–9.

Jàrvinen, A. and Vaisanen, R. A. (1983). Egg size and related reproductive traits in a southern passerine *Ficedula hypoleuca* breeding in an extreme northern environment. *Ornis Scand.*, 14, 253–62.

Jeanniard-du-Dot, T., Guinet, C., Arnould, J. P. Y., Speakman, J. R. and Trites, A. W. (2017). Accelerometers can measure total and activity-specific energy expenditures in free-ranging marine mammals only if linked to time–activity budgets. *Funct. Ecol.*, 31, 377–86.

Jensen, A. A. and Leffers, H. (2008). Emerging endocrine disrupters: perfluoroalkylated substances. *Int. J. Androl.*, 31, 161–9.

Johne, R. and Müller, H. (1998). Avian polyomavirus in wild birds: genome analysis of isolates from Falconiformes and Psittaciformes. *Arch. Virol.*, 143, 1501–12.

Johnstone, R. A. (1996). Multiple displays in animal communication: 'backup signals' and 'multiple messages'. *Phil. Trans. R. Soc. B*, 351, 329–38.

Jokimäki, J., Suhonen, J. and Kaisanlahti-Jokimäki, M.-L. (2018). Urban core areas are important for species conservation: a European-level analysis of breeding bird species. *Land. Urban. Plann.*, 178, 73–81.

Jones, C. G. (1984). Feeding ecology of the Mauritius kestrel. In J. Mendelsohn and C. W. Sapsford, eds., *Proceedings of the second symposium on African predatory birds* (p. 209). Durban, South Africa: Natal Bird Club.

Jones, C. G. (1987). The larger land birds of Mauritius. In A. W. Diamond, ed., *Studies of Mascarene island birds* (pp. 208–300). Cambridge: Cambridge University Press.

Jones, C. G. and Owadally, A. W. (1985). The status ecology and conservation of the Mauritius kestrel. In I. Newton and R. D. Chancellor, eds., *Conservation studies on raptors* (pp. 211–22). Cambridge: International Council for Bird Preservation.

Jones, C. G., Heck, W., Lewis, R. E., et al. (1995). The restoration of the Mauritius kestrel *Falco punctatus* population. *Ibis*, 137, 173–80.

Jones, C. G., Burgess, M. D., Groombridge, J. J., et al. (2013). Mauritius Kestrel *Falco punctatus*. In R. J. Safford and A. F. A. Hawkins, eds., *The birds of Africa. Volume VIII, The Malagasy region*. London: Christopher Helm.

Jönsson, K. I., Korpimäki, E., Pen, I. and Tolonen, P. (1996). Daily energy expenditure and short-term reproductive costs in free-ranging Eurasian kestrels (*Falco tinnunculus*). *Funct. Ecol.*, 10, 475–82.

Jönsson, K. I., Wiehn, J. and Korpimäki, E. (1999). Body reserves and unpredictable breeding conditions in the Eurasian kestrel, *Falco tinnunculus*. *Ecoscience*, 6, 406–14.

Kabouche, B. (1999). *Le niveau d'impact des lignes électriques Moyenne-Tension sur l'avifaune dans le secteur occidental du Var (St-Victoire–St-Baume–Mt-Faron)*. Rapport et carte CEEP – EDF services de distribution EDF Var.

Kaf, A., Saheb, M. and Bensaci, E. (2015). Preliminary data on breeding, habitat use and diet of common kestrel, *Falco tinnunculus*, in urban area in Algeria. Zool. *Ecol.*, 25, 203–10.

Kal'avský, M. and Pospíšilová, B. (2010). The ecology of ectoparasitic species *Carnus hemapterus* on nestlings of common kestrel (*Falco tinnunculus*) in Bratislava. *Slovak Raptor J.*, 4, 45–8.

Kangas, V.-M., Carrillo, J., Debray, P. and Kvist, L. (2018). Bottlenecks, remoteness and admixture shape genetic variation in island populations of Atlantic and Mediterranean common kestrels *Falco tinnunculus*. *J. Avian Biol.*, 49, e01768.

Karell, P., Ahola, K., Karstinen, T., Valkama, J. and Brommer, J. E. (2011). Climate change drives microevolution in a wild bird. *Nat. Comm.*, 2, 208.

Kay, S., Millet, J., Watson, J. and Shah, N. J. (2002). Status of the Seychelles kestrel Falco araea: a reassessment of the populations on Mahé and Praslin 2001–2002. Report by BirdLife Seychelles, Victoria, Mahé, Republic of Seychelles.

Kemp, A. C. (1995). A comparison of hunting behaviour by each sex of adult Greater Kestrels *Falco rupicoloides* resident near Pretoria, South Africa. *Ostrich*, 66, 21–33.

Kemp, A. C. (1999). Plumage development and visual communication in the Greater Kestrel *Falco rupicoloides* near Pretoria, South Africa. *Ostrich*, 70, 220–4.

Kettel, E. F., Gentle, L. K., Quinn, J. L. and Yarnell, R. W. (2018). The breeding performance of raptors in urban landscapes: a review and meta-analysis. *J. Ornithol.*, 159, 1–18.

Khaleghizadeh, A. and Javidkar, M. (2006). On the breeding season diet of the Common Kestrel, *Falco tinnunculus*, in Tehran, Iran. *Zool. Middle East*, 37, 113–4.

Kim, S.-Y., Fargallo, J. A., Vergara, P. and Martínez-Padilla, J. (2013). Multivariate heredity of melanin-based coloration, body mass and immunity. *Heredity*, 111, 139–46.

Kimball, R. T. (2006). Hormonal control of coloration. In G. E. Hill and K. J. McGraw, eds., *Bird coloration Vol. 1. Mechanisms and measurements* (pp. 431–68). Cambridge, MA: Harvard University Press.

King, A. J. and Cowlishaw, G. (2009). Foraging opportunities drive interspecific associations between rock kestrels and desert baboons. *J. Zool.*, 277, 111–8.

Kirkwood, J. K. (1979). The partition of food energy for existence in the kestrel (*Falco tinnunculus*) and the barn owl (*Tyto alba*). *Comp. Biochem. Physiol. Part A*, 63, 495–8.

Kitowski, I. (2005). Sex skewed kleptoparasitic exploitation of common kestrel *Falco tinnunculus*: the role of hunting costs to victims and tactics of kleptoparasites. *Folia Zool.*, 54, 371–8.

Kitzing, D. (1980). Neuere erkenntnisse ueber das falkenpockenvirus. *Der Praktische Tierazt.*, 61, 952–6.

Kochanek, H.-M. (1990). Ernährung des turmfalken (*Falco tinnunculus*): ergebnisse von nestinhaltsanalysen und automatischer registrierung. *J. Ornithol.*, 131, 291–304.

Koenig, A. (1890). Ornithologische forschungsergebnisse einer reise nach Madeira und den Canarischen Inseln. *J. Ornithol.*, 38, 257–488.

Koivula, M., Viitala, J. and Korpimäki, E. (1999). Kestrels prefer scent marks according to species and reproductive status of voles. *Ecoscience*, 6, 415–20.

Komen, J. and Myer, E. (1989). Observation on post-fledging dependence of kestrels (*Falco tinnunculus rupicolus*) in an urban environment. *J. Rapt. Res.*, 23, 94–8.

Korpimäki, E. (1985a). Diet of the kestrel *Falco tinnunculus* in the breeding season. *Ornis Fenn.*, 62, 130–7.

Korpimäki, E. (1985b). Prey choice strategies of the kestrel *Falco tinnunculus* in relation to available small mammals and other Finnish birds of prey. *Ann. Zool. Fenn.*, 22, 91–104.

Korpimäki, E. (1986). Diet variation, hunting habitat and reproductive output of the kestrel *Falco tinnunculus* in the light of the optimal diet theory. *Ornis Fenn.*, 63, 84–90.

Korpimäki, E. (1987). Dietary shifts, niche relationships and reproductive output of coexisting Kestrels and Long eared Owls. *Oecologia*, 74, 277–85.

Korpimäki, E. (1988). Factors promoting polygyny in European birds of prey – a hypothesis. *Oecologia*, 77, 278–85.

Korpimäki, E. (1994). Rapid or delayed tracking of multi-annual vole cycles by avian predators? *J. Anim. Ecol.*, 63, 619–28.

Korpimäki, E. and Norrdahl, K. (1991). Numerical and functional responses of kestrels, short-eared owls, and long-eared owls to vole densities. *Ecology*, 72, 814–26.

Korpimäki, E. and Rita, H. (1996). Effects of brood size manipulations on offspring and parental survival in the European kestrel under fluctuating food conditions. *Ecoscience*, 3, 264–73.

Korpimäki, E. and Wiehn, J. (1998). Clutch size of kestrels: seasonal decline and experimental evidence for food limitation under fluctuating food conditions. *Oikos*, 83, 259–72.

Korpimäki, E., Tolonen, P. and Valkama, J. (1994). Functional responses and load-size effect in central place foragers: data from the kestrel and some general comments. *Oikos*, 69, 504–10.

Korpimäki, E., Tolonen, P. and Bennett, G. F. (1995). Blood parasites, sexual selection and reproductive success of European kestrels. *Écoscience*, 2, 335–43.

Korpimäki, E., Lahti, K., May, C. A., et al. (1996). Copulatory behaviour and paternity determined by DNA fingerprinting in kestrels: effects of cyclic food abundance. *Anim. Behav.*, 51, 945–55.

Korpimäki, E., May, C. A., Parkin, D. T., Wetton, J. H. and Wiehn, J. (2000). Environmental- and parental condition-related variation in sex ratio of kestrel broods. *J. Avian Biol.*, 31, 128–34.

Korpimäki, E., Oksanen, L., Oksanen, T., et al. (2005). Vole cycles and predation in temperate and boreal zones of Europe. *J. Anim. Ecol.*, 74, 1150–9.

Kostrzewa, A. (1991). Interspecific interference competition in three European raptor species. *Ethol. Ecol. Evol.*, 3, 127–43.

Kostrzewa, R. (1989). *Achtjährige Untersuchungen zur Brutbiologie und Ökologie der Turmfalken Falco tinnunculus in der Niederrheinischen Bucht im Vergleich mit verschiedenen Gebieten in der Bundesrepublik Deutschland und Wets-Berlin.* PhD thesis, Universität Köln, Köln, Germany.

Kostrzewa, R. and Kostrzewa, A. (1990). Relationship of spring and summer weather to density and breeding performance of the common buzzard *Buteo buteo*, goshawk *Accipiter gentilis* and kestrel *Falco tinnunculus*. *Ibis*, 132, 550–8.

Kostrzewa, R. and Kostrzewa, A. (1991). Winter weather, spring and summer density, and subsequent breeding success of Eurasian kestrels, common buzzards, and northern goshawks. *Auk*, 108, 342–7.

Kostrzewa, R. and Kostrzewa, A. (1997). Der bruterfolg des turmfalken *Falco tinnunculus* in Deutchland: Ergebnisse 1985–1994. *J. Ornithol.*, 138, 73–82.

Kraaijeveld, K., Kraaijeveld-Smit, F. J. L. and Komdeur, J. (2007). The evolution of mutual ornamentation. *Anim. Behav.*, 74, 657–77.

Kreiderits, A., Gamauf, A., Krenn, H. W. and Sumasgutner, P. (2016). Investigating the influence of local weather conditions and alternative prey composition on the breeding performance of urban Eurasian Kestrels *Falco tinnunculus*. *Bird Study*, 63, 369–79.

Krueger, T. E. Jr. (1998). The use of electrical transmission pylons as nesting sites by the kestrel *Falco tinnunculus* in North-East Italy. In R. C. Chancellor, B.-U. Meyburg and J. J. Ferrero, eds., *Holarctic birds of prey* (pp. 141–8). Berlin: The World Working Group on Birds of Prey and Owls.

Kübler, S., Kupko, S. and Zeller, U. (2005). The kestrel (*Falco tinnunculus*) in Berlin: investigation of breeding biology and feeding ecology. *J. Ornithol.*, 146, 271–8.

Kurth, D. (1970). Der turmfalke (*Falco tinnunculus*) im Münchener Stadtgebiet. *Anz. Orn. Ges. Bayern*, 9, 2–12.

Kuusela, S. (1983). Breeding success of the Kestrel *Falco tinnunculus* in different habitats in Finland. *Proc. Third Nordic Congress Ornithol.*, 1981, 53–8.

Laaksonen, T., Lyytinen, S. and Korpimäki, E. (2004a). Sex-specific recruitment and brood sex ratios of Eurasian kestrels in a seasonally and annually fluctuating northern environment. *Evol. Ecol.*, 18, 215–30.

Laaksonen, T., Fargallo, J. A., Korpimäki, E., et al. (2004b). Year- and sex-dependent effects of experimental brood sex ratio manipulation on fledging condition of Eurasian kestrels. *J. Anim. Ecol.*, 73, 342–52.

Laaksonen, T., Ahola, M., Eeva, T., Väisänen, R. A. and Lehikoinen, E. (2006). Climate change, migratory connectivity and changes in laying date and clutch size of the pied flycatcher. *Oikos*, 114, 277–90.

Laaksonen, T., Negro, J. J., Lyytinen, S., et al. (2008). Effects of experimental brood size manipulation and gender on carotenoid levels of Eurasian kestrels *Falco tinnunculus*. *PLoS ONE*, 3, e2374.

Lack, D. (1947). The significance of clutch size. *Ibis*, 89, 302–52.

Lack, D. (1954). *The natural regulation of animal numbers*. London: Oxford University Press.

Lambrechts, M. M., Wiebe, K. L., Sunde, P., et al. (2012). Nest box design for the study of diurnal raptors and owls is still an overlooked point in ecological, evolutionary and conservation studies: a review. *J. Ornithol.*, 153, 23–34.

Leaver, D. (1951). Autumn behaviour of kestrels. *Brit. Birds*, 44, 27–8.

Lee, K. P., Simpson, S. J. and Wilson, K. (2008). Dietary protein-quality influences melanization and immune function in an insect. *Funct. Ecol.*, 22, 1052–61.

Lemus, J. A., Fargallo, J. A., Vergara, P., Parejo, D. and Banda, E. (2010). Natural cross chlamydial infection between livestock and free-living bird species. *PLoS ONE*, 5, e13512.

Li, Z., Zhou, Z., Deng, D., Li, Q. and Clarke, J. A. (2014). A falconid from the Late Miocene of Northwestern China yields further evidence of transition in Late Neogene steppe communities. *The Auk*, 131, 335–50.

Lihu, X., Jianjian, L., Chunfu, T. and Wenshan, H. (2007). Foraging area and hunting technique selection of common kestrel (*Falco tinnunculus*) in winter: the role of perch sites. *Acta Ecol. Sinica*, 27, 2160–6.

Liminana, R., Romero, M., Mellone, U. and Urios, V. (2012). Mapping the migratory routes and wintering areas of lesser kestrels *Falco naumanni*: new insights from satellite telemetry. *Ibis*, 154, 389–99.

Lin, J. Y. and Fisher, D. E. (2007). Melanocyte biology and skin pigmentation. *Nature*, 445, 843–50.

Lind, O., Mitkus, M., Olsson, P. and Kelber, A. (2013). Ultraviolet sensitivity and colour vision in raptor foraging. *J. Exp. Biol.*, 216, 1819–26.

Londei, T. (2002). The fox kestrel (*Falco alopex*) hovers. *J. Rapt. Res.*, 36, 236–7.

López-Idiáquez, D., Vergara, P., Fargallo, J. A. and Martinez-Padilla, J. (2016a). Female plumage coloration signals status to conspecifics. *Anim. Behav.*, 121, 101–6.

López-Idiáquez, D., Vergara, P., Fargallo, J. A. and Martinez-Padilla, J. (2016b). Old males reduce melanin-pigmented traits and increase reproductive outcome under worse environmental conditions in common kestrels. *Ecol. Evol.*, 6, 1224–35.

López-Idiáquez, D., Vergara, P., Fargallo, J. A. and Martinez-Padilla, J. (2018). Providing longer post-fledging periods increases offspring survival at the expense of future fecundity. *PLoS ONE*, 13, e0203152.

López-Idiáquez, D., Fargallo, J. A., López-Rull, I. and Martínez-Padilla, J. (2019). Plumage coloration and personality in early life: sexual differences in signalling. *Ibis*, 161, 216–21.

López-Rull, I., Vergara, P., Martínez-Padilla, J. and Fargallo, J. A. (2016). Early constraints in sexual dimorphism: survival benefits of feminized phenotypes. *J. Evol. Biol.*, 29, 231–40.

Loreau, M., Daufresne, T., Gonzalez, A., et al. (2013). Unifying sources and sinks in ecology and Earth sciences. *Biol. Rev.*, 88, 365–79.

Love, O. P., Bird, D. M. and Shutt, L. J. (2003a). Corticosterone levels during post-natal development in captive American kestrels (*Falco sparverius*). *Gen. Comp. Endocrinol.*, 130, 135–41.

Love, O. P., Shutt, L. J., Silfies, J. S., et al. (2003b). Effects of dietary PCB exposure on adrenocortical function in captive American kestrels (*Falco sparverius*). *Ecotoxicol.*, 12, 199–208.

Lozano, G. A. (1994). Carotenoids, parasites, and sexual selection. *Oikos*, 70, 309–11.

Lozano, G. A. (2009). Multiple cues in mate selection: the sexual interference hypothesis. *BioSci. Hypoth.*, 2, 37–42.

MacDonald, M. A. (1973). Bigamy in kestrel. *British Birds*, 66, 77–8.

Madroño, A., González, C. and Atienza, J. C. (2004). *Libro rojo de las aves de España*. Madrid: Dirección General da Biodiversidad, Ministerio de Medio Ambiente, SEO/BirdLife.

Malher, F. and Lesaffre, G. (2007). L'histoire des oiseaux nicheurs à Paris. *Alauda*, 75, 309–18.

Malher, F., Lesaffre, G., Zucca, M. and Coatmeur, J. (2010). *Oiseaux nicheurs de Paris. Un atlas urbain*. Paris: Delachauxet and Niestlé.

Manganaro, A., Ranazzi, L., Ranazzi, R. and Sorace, A. (1990). La dieta dell'allocco, *Strix aluco*, nel parco di Villa Doria Pamphili (Roma). *Riv. Ital. Orn.*, 60, 37–52.

Marasco, V. and Costantini, D. (2016). Signaling in a polluted world: oxidative stress as an overlooked mechanism linking contaminants to animal communication. *Front. Ecol. Evol.*, 4, 95.

Marteinson, S. C., Palace, V., Letcher, R. J. and Fernie, K. J. (2017) Disruption of thyroxine and sex hormones by 1,2-dibromo-4-(1,2-dibromoethyl)cyclohexane (DBE-DBCH) in American kestrels (*Falco sparverius*) and associations with reproductive and behavioral changes. *Environ. Res.*, 154, 389–97.

Martin, L. B., Hawley, D. M. and Ardia, D. R. (2011). An introduction to ecological immunology. *Funct. Ecol.*, 25, 1–4.

Martínez, J. E. and Calvo, J. F. (2006). *Rapaces diurnas y nocturnas de la Región de Murcia*. Serie técnica 1/06. Dirección General del Medio Natural. Murcia: Consejería de Industria y Medio Ambiente.

Martínez-Padilla, J. (2006). Prelaying maternal condition modifies the association between egg mass and T cell-mediated immunity in kestrels. *Behav. Ecol. Sociobiol.*, 60, 510–5.

Martínez-Padilla, J. and Viñuela, J. (2011). Hatching asynchrony and brood reduction influence immune response in common kestrel *Falco tinnunculus* nestlings. *Ibis*, 153, 601–10.

Martínez-Padilla, J., Martínez, J., Dávila, J. A., et al. (2004). Within-brood size differences, sex and parasites determine blood stress protein levels in Eurasian kestrel nestlings. *Funct. Ecol.*, 18, 426–34.

Martínez-Padilla, J., Dixon, H., Vergara, P., Pérez-Rodríguez, L. and Fargallo, J. A. (2010). Does egg colouration reflect male condition in birds? *Naturwissenschaften*, 97, 469–77.

Martínez-Padilla, J., Vergara, P. and Fargallo, J. A. (2017a). Increased lifetime reproductive success of first-hatched siblings in common kestrels *Falco tinnunculus*. *Ibis*, 159, 803–11.

Martínez-Padilla, J., López-Idiáquez, D., López-Perea, J. J., et al. (2017b). A negative association between bromadiolone exposure and nestling body condition in common kestrels: management implications for vole outbreaks. *Pest Manag. Sci.*, 73, 364–70.

Masman, D. (1986). *The annual cycle of the kestrel* Falco tinnunculus: *a study in behavioural energetics*. PhD thesis, University of Groningen, Groningen, the Netherlands.

Masman, D., Gordijn, M., Daan, S. and Dijkstra, C. (1986). Ecological energetics of the kestrel: field estimates of energy intake throughout the year. *Ardea*, 74, 24–39.

Masman, D., Daan, S. and Beldhuis, H. J. A. (1988a). Ecological energetics of the kestrel: daily energy expenditure throughout the year based on time-energy budget, food intake and doubly labeled water methods. *Ardea*, 76, 64–81.

Masman, D., Daan, S. and Dijkstra, C. (1988b). Time allocation in the kestrel (*Falco tinnunculus*), and the principle of energy minimization. *J. Anim. Ecol.*, 57, 411–32.

Masman, D., Dijkstra, C., Daan, S. and Bult, A. (1989). Energetic limitation of an avian parental effort: field experiments in the kestrel (*Falco tinnunculus*). *J. Evol. Biol.*, 2, 435–55.

Massemin, S., Korpimäki, E., Pöyri, V. and Zorn, T. (2002). Influence of hatching order on growth rate and resting metabolism of kestrel nestlings. *J. Avian Biol.*, 33, 235–44.

Massemin, S., Korpimäki, E., Zorn, T., Pöyri, V. and Speakman, J. R. (2003). Nestling energy expenditure of Eurasian kestrels *Falco tinnunculus* in relation to food intake and hatching order. *Avian Sci.*, 3, 1–12.

Mattson, M. P. and Calabrese, E. J. (2010). *Hormesis: a revolution in biology, toxicology and medicine*. New York: Springer.

McClure, C. J. W., Schulwitz, S. E., van Buskirk, R., Pauli, B. P. and Heath, J. A. (2017). Commentary: research recommendations for understanding the decline of American kestrels (*Falco sparverius*) across much of North America. *J. Raptor Res.*, 51, 455–64.

McDonald, P., Olsen, P. D. and Cockburn, A. (2004). Weather dictates reproductive success and survival in the Australian brown falcon *Falco berigora. J. Anim. Ecol.*, 73, 683–92.

McEwen, B. S. and Stellar, E. (1993). Stress and the individual – mechanisms leading to disease. *Arch. Int. Med.*, 153, 2093–101.

McGraw, K. J., Correa, S. M. and Adkins-Regan, E. (2006). Testosterone upregulates lipoprotein status to control sexual attractiveness in a colorful songbird. *Behav. Ecol. Sociobiol.*, 60, 117–22.

McKelvey, D. S. (1977). The Mauritius kestrel some notes on its breeding biology behaviour and survival potential. *Hawk Trust Annual Report*, 8, 19–21.

Mebs, T. and Schmidt, D. (2006). *Die Greifvögel Europas, Nordafrikas und Vorderasiens.* Stuttgart: Kosmos Verlag.

Meijer, T. (1988). *Reproductive decisions in the kestrel Falco tinnunculus.* PhD thesis, University of Groningen, Groningen, the Netherlands.

Meijer, T. (1989). Photoperiodic control of reproduction and molt in the kestrel, *Falco tinnunculus. J. Biol. Rhyth.*, 4, 351–64.

Meijer, T. and Schwabl, H. (1989). Hormonal patterns in breeding and non-breeding kestrels: field and laboratory studies. *Gen. Comp. Endocrinol.*, 74, 148–60.

Meijer, T., Daan, S. and Dijkstra, C. (1988). Female condition and reproductive decisions: the effect of food manipulations in free-living and captive kestrels. *Ardea*, 76, 141–54.

Meijer, T., Masman, D. and Daan, S. (1989). Energetics of reproduction in female kestrels. *Auk*, 106, 549–59.

Meijer, T., Daan, S. and Hall, M. (1990). Family planning in the kestrel (*Falco tinnunculus*): the proximate control of covariation of laying date and clutch size. *Behaviour*, 114, 117–36.

Meijer, T., Deerenberg, C., Daan, S. and Dijkstra, C. (1992). Egg-laying and photorefractoriness in the European Kestrel *Falco tinnunculus. Ornis Scand.*, 23, 405–10.

Metcalfe, N. B. and Alonso-Alvarez, C. (2010). Oxidative stress as a life-history constraint: the role of reactive oxygen species in shaping phenotypes from conception to death. *Funct. Ecol.*, 24, 984–96.

Metcalfe, N. B. and Monaghan, P. (2001). Compensation for a bad start: grow now, pay later? *Trends Ecol. Evol.*, 16, 254–60.

Meyer, R. L. and Balgooyen, T. G. (1987). A study and implications of habitat separation by sex of wintering American kestrels (*Falco sparverius*). In D. M. Bird and R. Bowman, eds., *Ancestral kestrel* (pp. 107–23). Ste. Anne de Bellevue, Quebec: Raptor Research Foundation and MacDonald Raptor Research Centre of McGill University.

Mikeš, V. (2003). *Feeding behaviour of urban and farmland kestrels* (Falco tinnunculus). BSc thesis, Faculty of Biological Sciences, University of South Bohemia, České Budějovice.

Mikula, P., Hromada, M., and Tryjanowski, P. (2013). Bats and swifts as food of the European kestrel (*Falco tinnunculus*) in a small town in Slovakia. *Ornis Fenn.*, 90, 178–85.

Miller, M. P., Mullins, T. D., Parrish, J. W. Jr, Walters, J. R. and Haig, S. M. (2012). Variation in migratory behavior influences regional genetic diversity and structure among American kestrel populations (*Falco sparverius*) in North America. *J. Hered.*, 103, 503–14.

Millon, A. and Bretagnolle, V. (2004). Les populations nicheuses de rapaces en France: analyse des résultats de l'enquête Rapaces 2000. In J. M. Thiollay and V. Bretagnolle, eds., *Rapaces nicheurs de France* (pp. 129–40). Paris: Delachaux and Niestlé.

Mock, D. W. and Parker, G. A. (1997). *The evolution of sibling rivalry.* Oxford, UK: Oxford University Press.

Mohammed, A. H. H. (1958). *Systematic and experimental studies on protozoal blood parasites of Egyptian birds. Part 1.* Cairo: Cairo University Press.

Møller, A. P. (1994). Facts and artefacts in nest box studies: implications for studies of birds of prey. *J. Raptor Res.*, 28, 143–8.

Møller, A. P. and Pomiankoski, A. (1993). Why have birds got multiple sexual ornaments? *Behav. Ecol. Sociobiol.*, 32, 167–76.

Møller, A. P., Sorci, G. and Erritzøe, J. (1998). Sexual dimorphism in immune defense. *Am. Nat.*, 152, 605–19.

Møller, A. P., Biard, C., Blount, J. D., et al. (2000). Carotenoid-dependent signals: indicators of foraging efficiency, immunocompetence or detoxification ability? *Avian Poult. Biol. Rev.*, 11, 137–59.

Montier, D. (1977). *Atlas of breeding birds of the London area.* London: Batsford.

Moreau, R. E. (1972). *The Palaearctic–African bird migration systems.* London: Academic Press.

Moreno, J. and Osorno, J. L. (2003). Avian egg colour and sexual selection: does eggshell pigmentation reflect female condition and genetic quality? *Ecol. Lett.*, 6, 803–6.

Morganti, M., Franzoi, A., Bontempo, L. and Sarà, M. (2016). An exploration of isotopic variability in feathers and claws of lesser kestrel *Falco naumanni* chicks from southern Sicily. *Avocetta*, 40, 23–32.

Morinha, F., Travassos, P., Seixas, F., et al. (2014) Differential mortality of birds killed at wind farms in Northern Portugal. *Bird Study*, 61, 255–9.

Motti, C., Leshem, Y., Izhaki, I. and Halevi, S. (2008). A case of polygamy or co-operative breeding in the common kestrel *Falco tinnunculus* in Israel. *Sandgrouse*, 30, 164–5.

Mougeot, F., Perez-Rodriguez, L., Martinez-Padilla, J., Leckie, F. and Redpath, S. M. (2007). Parasites testosterone and honest carotenoid-based signalling of health. *Funct. Ecol.*, 21, 886–98.

Mueller, H. C. (1971). Oddity and specific searching image more important than conspicuousness in prey selection. *Nature*, 233, 345–6.

Mueller, H. C. (1973). The relationship of hunger to predatory behaviour in hawks (*Falco sparverius* and *Buteo platypterus*). *Anim. Behav.*, 21, 513–20.

Mueller, H. C. (1974). Food caching behaviour in the American kestrel. *Z. Tierphsychol.*, 34, 105–14.

Mueller, H. C. (1987). Prey selection by kestrels: a review. In D. M. Bird and R. Bowman, eds., *Ancestral kestrel* (pp. 83–106). Ste. Anne de Bellevue, Quebec: Raptor Research Foundation and MacDonald Raptor Research Centre of McGill University.

Muir, D. C. and de Wit, C. A. (2010). Trends of legacy and new persistent organic pollutants in the circumpolar arctic: overview, conclusions, and recommendations. *Sci. Total Environ.*, 408, 3044–51.

Müller, C., Jenni-Eiermann, S. and Jenni, L. (2009). Effects of a short period of elevated circulating corticosterone on postnatal growth in free-living Eurasian kestrels *Falco tinnunculus*. *J. Exp. Biol.*, 212, 1405–12.

Müller, C., Jenni-Eiermann, S. and Jenni, L. (2010). Development of the adrenocortical response to stress in Eurasian kestrel nestlings: defence ability, age, brood hierarchy and condition. *Gen. Comp. Endocrinol.*, 168, 474–83.

Müller, C., Jenni-Eiermann, S. and Jenni, L. (2011). Heterophils/lymphocytes-ratio and circulating corticosterone do not indicate the same stress imposed on Eurasian kestrel nestlings. *Funct. Ecol.*, 25, 566–76.

Munck, A., Guyre, P. M. and Holbrook, N. J. (1984). Physiological functions of glucocorticoids in stress and their relation to pharmacological actions. *End. Rev.*, 5, 25–44.

Muñoz, E., Molina, R. and Ferrer, D. (1999). *Babesia shortti* infection in a common kestrel (*Falco tinnunculus*) in Catalonia (northeast Spain). *Avian Pathol.*, 28, 207–9.

Nagy, G. G., Ladányi, M., Arany, I., Aszalós, R. and Czúcz, B. (2017). Birds and plants: comparing biodiversity indicators in eight lowland agricultural mosaic landscapes in Hungary. *Ecol. Indic.* 73, 566–73.

Nantón, D. P. (2011). Factors determining the large-scale seasonal abundance of the common kestrel in Central Spain. *Ardeola*, 58, 87–101.

Nathan, R., Spiegel, O., Fortmann-Roe, S., et al. (2012). Using tri-axial acceleration data to identify behavioral modes of free-ranging animals: general concepts and tools illustrated for griffon vultures. *J. Exp. Biol.*, 215, 986–96.

Navarro-López, J. and Fargallo, J. A. (2015). Trophic niche in a raptor species: the relationship between diet diversity, habitat diversity and territory quality. *PLoS ONE*, 10, e0128855.

Navarro-López, J., Vergara, P. and Fargallo, J. A. (2014). Trophic niche width, offspring condition and immunity in a raptor species. *Oecologia*, 174, 1215–24.

Negro, J. J. and Grande, J. M. (2001). Territorial signalling: a new hypothesis to explain frequent copulation in raptorial birds. *Anim. Behav.*, 62, 803–9.

Negro, J. J., Donázar, J. A. and Hiraldo, F. (1992). Copulatory behaviour in a colony of lesser kestrels: sperm competition and mixed reproductive strategies. *Anim. Behav.*, 43, 921–30.

Negro, J. J., Villarroel, M. R., Tella, J. L., et al. (1996). DNA fingerprinting reveals a low incidence of extra-pair fertilizations and intraspecific brood parasitism in the lesser kestrel. *Anim. Behav.*, 51, 935–43.

Negro, J. J., Bortolotti, G. R., Tella, J. L., Fernie, K. J. and Bird, D. M. (1998). Regulation of integumentary colour and plasma carotenoids in American kestrels consistent with sexual selection theory. *Funct. Ecol.*, 12, 307–12.

Newton, E. (1867). On the land birds of the Seychelles archipelago. *Ibis*, 3, 335–60.

Newton, I. (1979). *Population ecology of raptors*. Berkhamsted: T. and A. D. Poyser.

Newton, I. (1998). *Population limitation in birds*. London: Academic Press.

Newton, I. (2013). Organochlorine pesticides and birds. *Brit. Birds*, 106, 189–205.

Newton, I. and Marquiss, M. (1981). Effect of additional food on laying dates and clutch sizes of sparrowhawks. *Ornis Scandin.*, 12, 224–9.

Newton, I., Wyllie, I. and Dale, L. (1999). Trends in the numbers and mortality patterns of sparrowhawks (*Accipiter nisus*) and kestrels (*Falco tinnunculus*) in Britain, as revealed by carcass analyses. *J. Zool.*, 248, 139–47.

Nichols, R. A., Bruford, M. W. and Groombridge, J. J. (2001). Sustaining genetic variation in a small population: evidence from the Mauritius kestrel. *Mol. Ecol.*, 10, 593–602.

Nicoll, M. A. C., Jones, C. G. and Norris, K. (2006). The impact of harvesting on a formerly endangered tropical bird: insight from life-history theory. *J. Appl. Ecol.*, 43, 567–75.

Noer, H. and Secher, H. (1983). Survival of Danish Kestrels *Falco tinnunculus* in relation to protection of birds of prey. *Ornis Scand.*, 14, 104–14.

Nore, T. (1979). Rapaces diurnes communs en Limousin pendant la période de nidification (II: Autour, Epervier et Faucon crécerelle). *Alauda*, 47, 259–69.

Olendorff, R. R. and Stoddart, J. W. Jr. (1974). The potential for management of raptor populations in western grassland. In F. N. Hamerstrom, B. E. Harrell and R. R. Olendorff, eds., *Management of raptors* (pp. 105–17). Vermillion, SD: Raptor Research Foundation, Inc. Raptor Report, 2.

Olsen, P. (1995). *Australian birds of prey: the biology and ecology of raptors*. Baltimore: John Hopkins University Press.

Olsen, P. and Baker, G. B. (2001). Daytime incubation temperatures in nests of the Nankeen Kestrel, *Falco cenchroides*. *Emu*, 101, 255–8.

Olsen, P. and Olsen, J. (1980). Observations on development, nesting chronology, and clutch and brood size in the Australian kestrel, *Falco cenchroides* (Aves: Falconidae). *Aust. Wildl. Res.*, 7, 247–55.

Olsen, P. and Olsen, J. (1987). Egg weight loss during incubation in captive Australian kestrels *Falco cenchroides* and brown goshawks *Accipiter fasciatus*. *Emu*, 87, 196–9.

Olsen, P. D., Marshall, R. C. and Gaal, A. (1989). Relationships within the genus *Falco*: a comparison of the electrophoretic patterns of feather proteins. *Emu*, 89, 193–203.

Olson, V. A. and Owens, I. P. F. (2005). Interspecific variation in the use of carotenoid-based coloration in birds: diet, life history and phylogeny. *J. Evol. Biol.*, 18, 1534–46.

Ozeki, H., Ito, S., Wakamatsu, K. and Ishiguro, I. (1997). Chemical characterization of pheo-melanogenesis starting from dihydroxyphenylalanine or tyrosine and cysteine. Effects of tyrosinase and cysteine concentrations and reaction time. *Biochim. Biophys. Acta*, 1336, 539–48.

Padilla, D. P. and Nogales, M. (2009). Behavior of kestrels feeding on frugivorous lizards: implications for secondary seed dispersal. *Behav. Ecol.*, 20, 872–7.

Padilla, D. P., Nogales, M. and Marrero, P. (2007). Prey size selection of insular lizards by two sympatric predatory bird species. *Acta Ornithol.*, 42, 167–72.

Padilla, D. P., González-Castro, A. and Nogales, M. (2012). Significance and extent of secondary seed dispersal by predatory birds on oceanic islands: the case of the Canary archipelago. *J. Anim. Ecol.*, 100, 416–27.

Palmer, R. S. (1988). *Handbook of North American birds*. Vol. 5, New Haven, CT: Yale University Press.

Palokangas, P., Alatalo, R. V. and Korpimäki, E. (1992). Female choice in the kestrel under different availability of mating options. *Anim. Behav.*, 43, 659–65.

Palokangas, P., Korpimäki, E., Hakkarainen, H., et al. (1994). Female kestrels gain reproductive success by choosing brightly ornamented males. *Anim. Behav.*, 47, 443–8.

Palozza, P. (1998). Prooxidant actions of carotenoids in biological systems. *Nutr. Rev.*, 56, 257–65.

Pang, S., Lee, P. P. N. and Tang, P. (1991). Sensory receptors as a special class of hormonal cells. *Neuroendocrinology*, 53, 2–11.

Parejo, D. and Silva, N. (2009a). Immunity and fitness in a wild population of Eurasian kestrels *Falco tinnunculus*. *Naturwissenschaften*, 96, 1193–202.

Parejo, D. and Silva, N. (2009b). Methionine supplementation influences melanin-based plumage colouration in Eurasian kestrel, *Falco tinnunculus*, nestlings. *J. Exp. Biol.*, 212, 3576–82.

Parejo, D., Silva, N., Danchin, E. and Avilés, J. M. (2011). Informative content of melanin-based plumage colour in adult Eurasian kestrels. *J. Avian Biol.*, 42, 49–60.

Parker, A. (1977). Kestrel hiding food. *Brit. Birds*, 70, 339–40.

Parr, D. (1969). A review of the status of the kestrel, tawny owl and barn owl in Surrey. *Surrey Bird Rep.*, 15, 35–42.

Paull, D. (1991). Foraging and breeding behaviour of the Australian Kestrel *Falco cenchroides* on the Northern Tablelands of New South Wales. Aust. *Bird Watcher*, 14, 85–92.

Peggie, C. T., Garratt, C. M. and Whittingham, M. J. (2011). Creating ephemeral resources: how long do the beneficial effects of grass cutting last for birds? *Bird Study*, 58, 390–8.

Peirce, M. A. (2000) A taxonomic review of avian piroplasms of the genus *Babesia* Starcovici, 1893 (Apicomplexa: Piroplasmorida: Babesiidae). *J. Nat. Hist.*, 34, 317–32.

Pen, I. (2000). *Sex allocation in a life history context*. PhD thesis, University of Groningen, Groningen, the Netherlands.

Pérez, C., Lores, M. and Velando, A. (2008) Availability of nonpigmentary antioxidant affects red coloration in gulls. *Behav. Ecol.*, 19, 967–73.

Peter, H.-U. and Zaumseil, J. (1982). Populations ökologische untersuchungen an einer Turmfalken kolonie *Falco tinnunculus* bei Jena. *Ber. Vogelwarte Hiddensee*, 3, 5–17.

Pettifor, R. A. (1983). Seasonal variation, and associated energetic implications, in the hunting behaviour of the kestrel. *Bird Study*, 30, 201–6.

Pettifor, R. A. (1984). Habitat utilisation and the prey taken by Kestrels in arable fenland. *Bird Study*, 31, 213–6.

Pettifor, R. A. (1990). The effects of avian mobbing on a potential predator, the European kestrel, *Falco tinnunculus. Anim. Behav.*, 39, 821–7.

Petty, S. J., Anderson, D. I. K., Davison, M., et al. (2003). The decline of Common Kestrels *Falco tinnunculus* in a forested area of northern England: the role of predation by Northern Goshawks *Accipiter gentilis. Ibis*, 145, 472–83.

Piattella, E., Salvati, L., Manganaro, A. and Fattorini, S. (1999). Spatial and temporal variations in the diet of the common kestrel (*Falco tinnunculus*) in urban Rome, Italy. *J. Rapt. Res.*, 33, 172–5.

Piechocki, R. (1982). *Der turmfalke*. Wittenberg: Lutherstadt Ziemsen Verlag.

Pikula, J., Beklová, M. and Kubík, V. (1984). The nidobiology of *Falco tinnunculus. Acta Sc. Nat. Brno*, 18, 1–55.

Plesník, J. (1990). Long-term study of some urban and extraurban populations of the kestrel (*Falco tinnunculus* L.). In K. Stastiny and V. Bejcek, eds., *Bird census and atlas studies* (pp. 453–8). Prague, Czech Republic: 11th International Conference on Bird Census and Atlas Work.

Plesník, J. and Dusík, M. (1994). Reproductive output of the Kestrel *Falco tinnunculus* in relation to small mammal dynamics in intensively cultivated farmland. In B.-U. Meyburg and R. C. Chancellor, eds., *Raptor conservation today* (pp. 61–5). Tonbridge: WWGBP and Pica Press.

Porter, R. D. (1975). Experimental alterations of clutch size of captive American kestrels *Falco sparverius. Ibis*, 117, 510–5.

Potgieter, L. N. D., Kocan, A. A. and Kocan, K. M. (1979). Isolation of a herpesvirus from American Kestrel with inclusion body disease. *J. Wildl. Dis.*, 15, 143–9.

Pranty, B., Kwater, E., Weatherman, H. and Robinson, H. P. (2004). Eurasian kestrel in Florida: first record for the south-eastern United States, with a review of its status in North America. *N. Am. Birds*, 58, 168–9.

Price, T., Kirkpatrick, M. and Arnold, S. J. (1988). Directional selection and the evolution of breeding date in birds. *Science*, 240, 798–9.

Prugh, L. R., Stoner, C. J., Epps, C. W., et al. (2009). The rise of the mesopredator. *Bioscience*, 59, 779–91.

Pulliam, H. R. (1988). Sources, sinks, and population regulation. *Am. Nat.*, 132, 652–61.

Purger, J. J. (1998). Diet of red-footed falcon *Falco vespertinus* nestlings from hatching to fledging. *Ornis Fenn.*, 75, 185–91.

Qasem, L., Cardew, A., Wilson, A., et al. (2012). Tri-axial dynamic acceleration as a proxy for animal energy expenditure; should we be summing values or calculating the vector? *PLoS ONE*, 7, e31187.

Quinn, M. J., French, J. B., McNabb, F. M. A. and Ottinger, M. A. (2002). The effects of polychlorinated biphenyls (Aroclor 1242) on thyroxine, estradiol, molt, and plumage characteristics in the American kestrel (*Falco sparverius*). *Environm. Toxicol. Chem.*, 21, 1417–22.

Raida, S. R. and Jaensch, S. M. (2000). Central nervous disease and blindness in Nankeen kestrels (*Falco cenchroides*) due to a novel *Leucocytozoon*-like infection. *Avian Pathol.*, 29, 51–6.

Rand, A. L. (1936). The distribution and habits of Madagascar birds: summary of the field notes of the Mission Zoologique Franco-Anglo-Américaine à Madagascar. *Bull. Amer. Mus. Nat. Hist.*, 72, 143–499.

Ratcliffe, D. A. (1967). Decrease in eggshell weight in certain birds of prey. *Nature*, 215, 208–10.

Ratcliffe, D. A. (1970). Changes attributable to pesticides in egg breakage frequency and eggshell thickness in some British birds. *J. Appl. Ecol.*, 7, 67–107.

Raubenheimer, D., Simpson, S. J. and Mayntz, D. (2009). Nutrition, ecology and nutritional ecology: toward an integrated framework. *Funct. Ecol.*, 23, 4–16.

Redpath, S. M., Arroyo, B. E., Etheridge, B., et al. (2002). Temperature and hen harrier productivity: from local mechanisms to geographical patterns. *Ecography*, 25, 533–40.

Regoli, F. and Giuliani, M. E. (2014). Oxidative pathways of chemical toxicity and oxidative stress biomarkers in marine organisms. *Mar. Environ. Res.*, 93, 106–17.

Rehder, N. B., Bird, D. M. and Lague, P. C. (1986). Variations in plasma corticosterone, estrone, estradiol-17β, and progesterone concentrations with forced renesting, molt, and body weight of captive female American kestrels. *Gen. Comp. Endocrinol.*, 62, 386–93.

Reiter, R. J., Tan, D. X., Cabrera, J., et al. (1999). The oxidant/antioxidant network: role of melatonin. *Biol. Signals Recep.*, 8, 56–63.

Rejt, L. 2001. Peregrine Falcon and Kestrel in urban environments – the case of Warsaw. In E. Gottschalk, A. Barkow, M. Mühlenberg and J. Settele, eds., *Naturschutz und verhalten* (pp. 81–5). Leipzig: UFZ-Bericht, UFZ Leipzig-Halle.

Rene de Roland, L.-A., Rabearivony, J., Razafimanjato, G., Robenarimangason, H. and Thorstrom, R. (2005a). Breeding biology and diet of Banded Kestrels *Falco zoniventris* on Masoala Peninsula, Madagascar. *Ostrich*, 76, 32–6.

Rene de Roland, L.-A., Rabearivony, J., Robenarimangason, H., Razafimanjato, G. and Thorstrom, R. (2005b). Breeding biology and diet of the Madagascar Kestrel (*Falco newtoni*) in Northeastern Madagascar. *J. Rapt. Res.*, 39, 149–55.

Reuter, G., Boros, Á., Mátics, R., et al. (2016). Divergent hepatitis E virus in birds of prey, common kestrel (*Falco tinnunculus*) and red-footed falcon (*F. vespertinus*), Hungary. *Infect. Genet. Evol.*, 43, 343–6.

Riddle, G. S. (1979). The kestrel in Ayrshire 1970–78. *Scott. Birds*, 10, 201–15.

Riddle, G. S. (1987). Variation in the breeding output of kestrel pairs in Ayrshire 1978–85. *Scott. Birds*, 14, 138–45.

Riddle, G. S. (1991). *Season with the kestrel*. Cassell and Co.

Riddle, G. S. (2007). Common kestrel. In R. W. Forrester, I. J. Andrews, C. J. McInerny, et al., eds., *The birds of Scotland* (pp. 493–6). Aberlady: The Scottish Ornithologists' Club.

Riegert, J. and Fuchs, R. (2011). Fidelity to roost sites and diet composition of wintering male urban common kestrels *Falco tinnunculus*. *Acta Ornithol.*, 46, 183–9.

Riegert, J., Fainová, D., Mikeš, V. and Fuchs, R. (2007a). How urban Kestrels *Falco tinnunculus* divide their hunting grounds: partitioning or cohabitation? *Acta Ornithol.*, 42, 69–76.

Riegert, J., Dufek, A., Fainová, D., Mikeš, V. and Fuchs, R. (2007b) Increased hunting effort buffers against vole scarcity in an urban kestrel *Falco tinnunculus* population. *Bird Study*, 54, 353–61.

Riegert, J., Fainová, D. and Bystřická, D. (2010). Genetic variability, body characteristics and reproductive parameters of neighbouring rural and urban common kestrel (*Falco tinnuculus*) populations. *Popul. Ecol.*, 52, 73–9.

Rijnsdorp, A., Daan, S. and Dijkstra, C. (1981). Hunting in the kestrel, *Falco tinnunculus*, and the adaptive significance of daily habits. *Oecologia*, 50, 391–406.

Robinson, M. R., Pilkington, J. G., Clutton-Brock, T. H., Pemberton, J. M. and Kruuk, L. E. B. (2008). Environmental heterogeneity generates fluctuating selection on a secondary sexual trait. *Curr. Biol.*, 18, 751–7.

Robinson, R. A., Morrison, C. A. and Baillie, S. R. (2014). Integrating demographic data: towards a framework for monitoring wildlife populations at large spatial scales. *Methods Ecol. Evol.*, 5, 1361–72.

Rockenbauch, D. (1968). Zur brutbiologie des turmfalken (*Falco tinnunculus* L.). *Anz. Orn. Ges. Bayern*, 8, 267–76.

Rodríguez, A., Negro, J. J., Mulero, M., et al. (2012). The eye in the sky: combined use of unmanned aerial systems and GPS data loggers for ecological research and conservation of small birds. *PLoS ONE*, 7, e50336.

Rodríguez, A., Broggi, J., Alcaide, M., Negro, J. J. and Figuerola, J. (2014). Determinants and short-term physiological consequences of PHA immune response in lesser kestrel nestlings. *J. Exp. Zool.*, 321, 376–86.

Rodríguez, B., Rodríguez, A., Siverio, F. and Siverio, M. (2018). Factors affecting the spatial distribution and breeding habitat of an insular cliff-nesting raptor community. *Curr. Zool.*, 64, 173–81.

Rodríguez, C. and Bustamante, J. (2003). The effect of weather on lesser kestrel breeding success: can climate change explain historical population declines? *J. Anim. Ecol.*, 72, 793–810.

Romero, L. M. (2004). Physiological stress in ecology: lessons from biomedical research. *Trends Ecol. Evol.*, 19, 249–55.

Romero, L. M., Dickens, M. J. and Cyr, N. E. (2009). The reactive scope model – a new model integrating homeostasis, allostasis, and stress. *Horm. Behav.*, 55, 375–89.

Rubino, F. M., Pitton, M., Brambilla, G. and Colombi, A. (2006). A study of the glutathione metaboloma peptides by energy-resolved mass spectrometry as a tool to investigate into the interference of toxic heavy metals with their metabolic processes. *J. Mass Spectrom.*, 41, 1578–93.

Rudolf, S. G. (1982). Foraging strategies of American kestrels during breeding. *Ecology*, 63, 1268–76.

Rutkowski, R., Rejt, Ł. and Szczuka, A. (2006). Analysis of microsatellite polymorphism and genetic differentiation in urban and rural kestrels *Falco tinnunculus* (L.). *Pol. J. Ecol.*, 54: 473–80.

Rutkowski, R., Rejt, Ł., Tereba, A., Gryczyńska-Siemiątkowska, A. and Janic, B. (2010). Population genetic structure of the European kestrel *Falco tinnunculus* in central Poland. *Eur. J. Wildl. Res.*, 56, 297–305.

Safford, R. J. and Jones, C. G. (1997). Did organochlorine pesticide use cause declines in Mauritian forest birds? *Biol. Cons.*, 6, 1445–51.

Sakamoto, K. Q., Sato, K., Ishizuka, M., et al. (2009). Can ethograms be automatically generated using body acceleration data from free ranging birds? *PLoS ONE*, 4, e5379.

Salvati, L. (2002). Spring weather and breeding success of the Eurasian Kestrel (*Falco tinnunculus*) in urban Rome, Italy. *J. Rapt. Res.*, 36, 81–4.

Salvati, L., Manganaro, A., Fattorini, S. and Piattella, E. (1999). Population features of kestrels *Falco tinnunculus* in urban, suburban and rural areas in central Italy. *Acta Ornithol.*, 34, 53–8.

Sapolsky, R. M., Romero, L. M. and Munck, A. U. (2000). How do glucocorticoids influence stress responses? Integrating permissive, suppressive, stimulatory, and preparative actions. *Endocrin. Rev.*, 21, 55–89.

Sarno, R. J. and Gubanich, A. A. (1995). Prey selection by wild American kestrels: the influence of prey size and activity. *J. Raptor Res.*, 29, 123–6.

Schifferli, A. (1965). Vom zugverhalten der in der schweiz brutenden turmfalken, *Falco tinnunculus*, nach den ringfunden. *Orn. Beob.*, 62, 1–13.

Schmid, H. (1990) Die bestandsentwicklung des turmfalken *Falco tinnunculus* in der Schweiz. *Ornithol. Beob.*, 87, 327–49.

Schmid, H., Burkhardt, M., Keller, V., et al. (2001). *Die entwicklung der vogelwelt in der Schweiz*. Avifauna Report Sempach 1, Annex. Schweiz: Vogelwarte Sempach.

Sebastian-Gonzalez, E., Perez-Garcia, J., Carrete, M., Donazar, J. and Sanchez-Zapata, J. (2018). Using network analysis to identify indicator species and reduce collision fatalities at wind farms. *Biol. Cons.*, 224, 209–12.

Senapathi, D., Nicoll, M. A., Teplitsky, C., Jones, C. G. and Norris, K. (2011). Climate change and the risks associated with delayed breeding in a tropical wild bird population. *Proc. R. Soc. Lond. B*, 278, 3184–90.

Shan, Y., Pepe, J., Lu, T. H., et al. (2000). Induction of the heme oxygenase-1 gene by metalloporphyrins. *Arch. Biochem. Biophys.*, 380, 219–27.

Shaw, G. and Riddle, G. S. (2003). Comparative responses of barn owls (*Tyto alba*) and kestrels (*Falco tinnunculus*) to vole cycles in South-West Scotland. In D. B. A. Thomson, S. M. Redpath, A. H. Fielding, M. Marquis and C. A. Galbraith, eds., *Birds of prey in a changing environment* (pp. 131–6). Edinburgh: The Stationery Office.

Sheldon, B. C. and Verhulst, S. (1996). Ecological immunology: costly parasite defences and trade-offs in evolutionary ecology. *Trends Ecol. Evol.*, 11, 317–21.

Shrubb, M. (1970). The present status of the Kestrel in Sussex. *Bird Study*, 17, 1–15.

Shrubb, M. (1980). Farming influences on the food and hunting of Kestrels. *Bird Study*, 27, 109–15.

Shrubb, M. (1982). The hunting behaviour of some farmland kestrels. *Bird Study*, 28, 121–8.

Shrubb, M. (1993). *The kestrel*. London: Hamlyn.

Shutt, L. and Bird, D. M. (1985). Influence of nestling experience on nest-type selection in captive kestrels. *Anim. Behav.*, 33, 1028–31.

Sibley, C. G. and Ahlquist, J. E. (1990). *Phylogeny and classification of birds. A study in molecular evolution*. New Haven, CT: Yale University Press.

Simons, M. J. P. (2013). *Sexual coloration and aging*. PhD thesis, University of Groningen, Groningen, the Netherlands.

Slagsvold, T., Sandvik, J., Rofstad, G., Lorensten, O. and Husby, M. (1984). On the adaptive significance of intra-clutch egg size variation in birds. *The Auk*, 101, 685–97.

Śliwa, P. and Rejt, Ł. (2006). *Pustułka*. Świebodzin: Wydawnictwo Klubu Przyrodników.

Śliwa, P., Mokwa, K. and Rejt, Ł. (2010). Migrations and wintering of the kestrel (*Falco tinnunculus*) in Poland. *Ring*, 31, 59–69.

Smallwood, J. A. (1989). Prey preferences of free-ranging American kestrels, *Falco sparvierus*. *Anim. Behav.*, 38, 712–4.

Smallwood, K. S. and Thelander, C. G. (2004). *Developing methods to reduce bird mortality in the Altamont Pass Wind Resource Area*. Final report to the California Energy Commission, PIER-EA contract No. 500–01-019, Sacramento, California, USA.

Smallwood, P. D. and Smallwood, J. A. (1998). Seasonal shifts in sex ratios of fledgling American kestrels (*Falco sparverius paulus*): the early bird hypothesis. *Evol. Ecol.*, 12, 839–53.

Smart, J. and Amar, A. (2018). Diversionary feeding as a means of reducing raptor predation at seabird breeding colonies. *J. Nat. Cons.*, 46, 48–55.

Smiddy, P. (2017). Diet of the common kestrel *Falco tinnunculus* in east Cork and west Waterford: an insight into the dynamics of invasive mammal species. *Biol. Envir. Proc. R. Irish Acad.*, 117, 131–8.

Smit, T., Eger, A., Haagsma, J. and Bakhuizen, T. (1987). Avian tuberculosis in wild birds in the Netherlands. *J. Wildl. Dis.*, 23, 485–7.

Smits, J. E., Fernie, K. J., Bortolotti, G. R. and Marchant, T. A. (2002). Thyroid hormone suppression and cell-mediated immunomodulation in American Kestrels (*Falco sparverius*) exposed to PCBs. *Arch. Environm. Contam. Toxicol.*, 43, 338–44.

Snow, D. W. (1968). Movements and mortality of British kestrels *Falco tinnunculus*. *Bird Study*, 15, 65–83.

Snow, D. W. (1978). *An atlas of speciation in African non-passerine birds*. London: British Museum of Natural History.

Sockman, K. W. and Schwabl, H. (2001). Plasma corticosterone in nestling American kestrels: effects of age, handling stress, yolk androgens, and body condition. *Gen. Comp. Endocrin.*, 122, 205–12.

Sockman, K. W., Schwabl, H. and Sharp, P. J. (2000). The role of prolactin in the regulation of clutch size and onset of incubation behavior in the American kestrel. *Horm. Behav.*, 38, 168–76.

Sommani, E. (1986). Note sulla biologia di alcune coppie di gheppio, *Falco tinnunculus*, presenti in Roma. *Riv. Ital. Orn.*, 56, 40–52.

Sonerud, G. A. (1992). Functional response of birds of prey: biases due to the load-size effect in central place foragers. *Oikos*, 63, 223–32.

Sonerud, G. A., Steen, R., Løw, L. M., et al. (2013). Size-biased allocation of prey from male to offspring via female: family conflicts, prey selection, and evolution of sexual size dimorphism in raptors. *Oecologia*, 172, 93–107.

Sorace, A., and Gustin, M. (2010). Bird species of conservation concern along urban gradients in Italy. *Biod. Cons.*, 19, 205–21.

Sorensen, M. C., Hipfner, J. M., Kyser, T. K. and Norris, D. R. (2009). Carry-over effects in a Pacific seabird: stable isotope evidence that pre-breeding diet quality influences reproductive success. *J. Anim. Ecol.*, 78, 460–7.

Soulé, M. E., Bolger, D. T., Alberts, A. C., et al. (1988). Reconstructed dynamics of rapid extinctions of Chaparral-requiring birds in urban habitat islands. *Cons. Biol.*, 2, 75–92.

Souttou, K., Baziz, B., Doumandji, S., Denys, C. and Brahimi, R. (2007). Prey selection in the common kestrel, *Falco tinnunculus* (Aves, Falconidae) in the Algiers suburbs (Algeria). *Folia Zool.*, 56, 405–15.

Souttou, K., Manaa, A., Baziz-Neffah, F., et al. (2018). Geographic variation of the diet of the common kestrel *Falco tinnunculus* Linné, 1758 (Aves, Falconidae) in Algeria. *Vie et Milieu – Life Environ.*, 68, 127–43.

Steen, R., Løw, L. M., Sonerud, G. A., Selås, V. and Slagsvold, T. (2010). The feeding constraint hypothesis: prey preparation as a function of nestling age and prey mass in the Eurasian kestrel. *Anim. Behav.*, 80, 147–53.

Steen, R., Løw, L. M., Sonerud, G. A., Selås, V. and Slagsvold, T. (2011). Prey delivery rates as estimates of prey consumption by Eurasian Kestrel *Falco tinnunculus* nestlings. *Ardea*, 99, 1–8.

Steenhof, K. and Peterson, B. E. (2009). Site fidelity, mate fidelity, and breeding dispersal in American kestrels. *Wilson J. Orn.*, 121, 12–21.

Steinhagen, P. and Schellhaas, G. (1968). Pasteurellose in einer Falknerei. *Berliner und Münchener Tierärtzliche Wochenschrift*, 81, 72–5.

Stokes, A. W. (1971). Parental and courtship feeding in the red junglefowl. *The Auk*, 88, 21–9.

Strasser, E. H. and Heath, J. A. (2013). Reproductive failure of a human-tolerant species, the American Kestrel, is associated with stress and human disturbance. *J. Appl. Ecol.*, 50, 912–9.

Sumasgutner, P., Schulze, C. H., Krenn, H. W. and Gamauf, A. (2014a). Conservation related conflicts in nest-site selection of the Eurasian kestrel (*Falco tinnunculus*) and the distribution of its avian prey. *Land. Urban Plann.*, 127, 94–103.

Sumasgutner, P., Vasko, V., Varjonen, R. and Korpimäki, E. (2014b). Public information revealed by pellets in nest sites is more important than ecto-parasite avoidance in the settlement decisions of Eurasian kestrels. *Behav. Ecol. Sociobiol.*, 68, 2023–34.

Sumasgutner, P., Nemeth, E., Tebb, G., Krenn, H. W. and Gamauf, A. (2014c). Hard times in the city – attractive nest sites but insufficient food supply lead to low reproduction rates in a bird of prey. *Front Zool.*, 11, 48.

Sumasgutner, P., Terraube, J., Coulon, A., et al. (2019). Landscape homogenization due to agricultural intensification disrupts the relationship between reproductive success and main prey abundance in an avian predator. *Front. Zool.*, 16, 31.

Surai, P. (2002). *Natural antioxidants in avian nutrition and reproduction*. Nottingham: Nottingham University Press.

Surai, P. F., Speake, B. K., Noble, R. C. and Sparks, N. H. C. (1999). Tissue-specific antioxidant profiles and susceptibility to lipid peroxidation of the newly hatched chick. *Biol. Trace Elem. Res.*, 68, 63–78.

Tapia, L., Regos, A., Gil-Carrera, A. and Domínguez, J. (2017). Unravelling the response of diurnal raptors to land use change in a highly dynamic landscape in northwestern Spain: an approach based on satellite earth observation data. *Eur. J. Wildl. Res.*, 63, 40.

Tella, J. L., Donazar, J. A., Negro, J. J. and Hiraldo, F. (1996a). Seasonal and interannual variation in the sex-ratio of lesser kestrel *Falco naumanni* broods. *Ibis*, 138, 342–5.

Tella, J. L., Hiraldo, F., Donázar-Sancho, J. A. and Negro, J. J. (1996b). Costs and benefits of urban nesting in the lesser kestrel. In D. M. Bird, D. E. Varland and J. J. Negro, eds., *Raptors in human landscapes: adaptations to built and cultivated environments* (pp. 53–60). London: Academic Press.

Tella, J. L., Forero, M. G., Hiraldo, F. and Donazar, J. A. (1998). Conflicts between lesser kestrel conservation and European agricultural policies as identified by habitat use analyses. *Cons. Biol.*, 12, 593–604.

Temple, S. A. (1977). The status and conservation of endemic kestrels on the Indian Ocean islands. In R. D. Chancellor, ed., *Proceedings of the World Conference on Birds of Prey, Vienna 1975* (pp. 74–83). London: International Council for Bird Preservation.

Temple, S. A. (1987). Foraging ecology of the Mauritius Kestrel (*Falco punctatus*). *Biotropica*, 19, 2–6.

Terraube, J. and Bretagnolle, V. (2018). Top-down limitation of mesopredators by avian top predators: a call for research on cascading effects at the community and ecosystem scale. *Ibis*, 160, 693–702.

Terraube, J., Vasko, V., and Korpimaki, E. (2015). Mechanisms and reproductive consequences of breeding dispersal in a specialist predator under temporally varying food conditions. *Oikos*, 124, 762–71.

Thiollay, J.-M. (2007). Raptor population decline in West Africa. *Ostrich*, 78, 405–13.

Thiollay, J.-M. and Bretagnolle, V. (2004). *Rapaces nicheurs de France: distribution, effectifs et conservation*. Paris: Delachaux and Niestlé.

Thomas, N. J., Hunter, D. B. and Atkinson, C. T. (2007). *Infectious diseases of wild birds*. Ames, IA: Blackwell Publishing.

Thomson, A. L. (1958). The migration of British falcons (Falconidae) as shown by ringing results. *British Birds*, 51, 179–189.

Tieszen, L. L. and Boutton, T. W. (1988). Stable carbon isotopes in terrestrial ecosystem research. In P. W. Rundel, J. R. Ehleringer and K. A. Nagy, eds., *Stable isotopes in ecological research (Ecological Studies 68)* (pp. 167–95). Berlin: Springer.

Tinbergen, L. (1940). Beobachtungen über die arbeitsteilung des turmfalken (*Falco tinnunculus*) während der fortpflanzungsziet. *Ardea*, 29, 63–98.

Tinbergen, L. (1960). The natural control of insects in pinewoods. 1: Factors influencing the intensity of predation by songbirds. *Arch. Néer. Zool.*, 13, 265–336.

Tingay, R. E. and Gilbert, M. (2000). Behaviour of Banded Kestrel in western Madagascar: a possible foraging association with Sickle-billed Vanga. *Bull. Afric. Bird Club*, 7, 111–3.

Toland, B. R. (1987). The effect of vegetative cover on foraging strategies, hunting success and nesting distribution of American kestrels in central Missouri. *J. Raptor Res.*, 21, 14–20.

Tolonen, P. and Korpimäki, E. (1994). Determinants of parental effort: a behavioural study in the Eurasian kestrel, *Falco tinnunculus*. *Behav. Ecol. Sociobiol.*, 35, 355–62.

Tolonen, P. and Korpimaki, E. (1995). Parental effort of kestrels (*Falco tinnunculus*) in nest defense: effects of laying time, brood size, and varying survival prospects of offspring. *Behav. Ecol.*, 6, 435–41.

Tolonen, P. and Korpimaki, E. (1996). Do kestrels adjust their parental effort to current or future benefit in a temporally varying environment? *Ecoscience*, 3, 165–72.

Umanskaja, A. S. (1981). The Miocene birds of the western Black Sea coasts of the Ukrainian SSR. *Vestnik Zoologii*, 17, 17–21. [In Russian with English summary.]

Ursua, E., Serrano, D. and Tella, J. L. (2005). Does land irrigation actually reduce foraging habitat for breeding lesser kestrels? The role of crop types. *Biol. Cons.*, 122, 643–8.

Valkama, J. and Korpimäki, E. (1999). Nestbox characteristics, habitat quality and reproductive success of Eurasian kestrels. *Bird Study*, 46, 81–8.

Valkama, J., Korpimäki, E. and Tolonen, P. (1995). Habitat utilization, diet and reproductive success in the kestrel in a temporally and spatially heterogeneous environment. *Ornis Fenn.*, 72, 49–61.

Valkama, J., Korpimäki, E., Wiehn, J. and Pakkanen, T. (2002). Inter-clutch egg size variation in kestrels *Falco tinnunculus*: seasonal decline under fluctuating food conditions. *J. Avian Biol.*, 33, 426–32.

van de Brink, V., Henry, I., Wakamatsu, K. and Roulin, A. (2012). Melanin-based coloration in juvenile kestrels (*Falco tinnunculus*) covaries with anti-predatory personality traits. *Ethology*, 118, 673–82.

van Helden, Y. G., Keijer, J., Knaapen, A. M., et al. (2009). Beta-carotene metabolites enhance inflammation-induced oxidative DNA damage in lung epithelial cells. *Free Radic. Biol. Med.*, 46, 299–304.

van Noordwijk, A. L., Van Balen, J. H. and Scharloo, W. (1981). Genetic variation in the timing of reproduction in the great tit. *Oecologia*, 49, 158–66.

van Zyl, A. J. (1994). A comparison of the diet of the common kestrel *Falco tinnunculus* in South Africa and Europe. *Bird Study*, 41, 124–30.

Vasko, V., Laaksonen, T., Valkama, J. and Korpimäki, E. (2011). Breeding dispersal of Eurasian kestrels *Falco tinnunculus* under temporally fluctuating food abundance. *J. Avian Biol.*, 42, 552–63.

Vaurie, C. (1961). Systematic notes on palearctic birds. No. 45. Falconidae: the Genus Falco (Part 2). *Am. Mus. Novit.*, 2038, 13–21.

Vergara, P. and Fargallo, J. A. (2007). Delayed plumage maturation in Eurasian kestrels: female mimicry, subordination signalling or both? *Anim. Behav.*, 74, 1505–13.

Vergara, P. and Fargallo, J. A. (2008a). Copulation duration during courtship predicts fertility in the Eurasian kestrel *Falco tinnunculus*. *Ardeola*, 55, 153–60.

Vergara, P. and Fargallo, J. A. (2008b). Sex melanic colouration and sibling competition during the post-fledging dependence period. *Behav. Ecol.*, 19, 847–53.

Vergara, P. and Fargallo, J. A. (2011). Multiple coloured ornaments in male common kestrels: different mechanisms to convey quality. *Naturwissenschaften*, 98, 289–98.

Vergara, P., De Neve, L. and Fargallo, J. A. (2007). Agonistic behaviour prior to laying predicts clutch size in Eurasian kestrels: an experiment with natural decoys. *Anim. Behav.*, 74, 1515–23.

Vergara, P., Fargallo, J. A. and Martínez-Padilla, J. (2009). Inter-annual variation and information content of melanin-based coloration in female Eurasian kestrels. *Biol. J. Linn. Soc.*, 97, 781–90.

Vergara, P., Fargallo, J. A. and Martínez-Padilla, J. (2010). Reaching independence: food supply, parent quality, and offspring phenotypic characters in kestrels. *Behav. Ecol.*, 21, 507–12.

Vergara, P., Martínez-Padilla, J. and Fargallo, J. A. (2013). Differential maturation of sexual traits: revealing sex while reducing male and female aggressiveness. *Behav. Ecol.*, 24, 237–44.

Vergara, P., Fargallo, J. A. and Martínez-Padilla, J. (2015). Genetic basis and fitness correlates of dynamic carotenoid-based ornamental coloration in male and female common kestrels *Falco tinnunculus*. *J. Evol. Biol.*, 28, 146–54.

Vesey-Fitzgerald, D. (1940). On the birds of the Seychelles 1 – the endemic birds (land birds). *Ibis*, 82, 480–9.

Videler, J. J., Weihs, D. and Daan, S. (1983). Intermittent gliding in the hunting flight of the kestrel *Falco tinnunculus* L. *J. Exp. Biol.*, 102, 1–12.

Viitala, J., Korpimäki, E., Palokangas, P. and Koivula, M. (1995). Attraction of common kestrels to vole scent marks visible in ultraviolet light. *Nature*, 373, 425–7.

Village, A. (1982a). The diet of Kestrels in relation to vole abundance. *Bird Study*, 29, 129–38.

Village, A. (1982b). The home range and density of kestrels in relation to vole abundance. *J. Anim. Ecol.*, 51, 413–28.

Village, A. (1983). The role of nest-site availability and territorial behaviour in limiting the breeding density of kestrels. *J. Anim. Ecol.*, 52, 635–45.

Village, A. (1984). Problems in estimating kestrel breeding density. *Bird Study*, 31, 121–5.

Village, A. (1985a). Spring arrival times and assortative mating of kestrels in South Scotland. *J. Anim. Ecol.*, 54, 857–68.

Village, A. (1985b). Turnover, age and sex ratios of kestrels (*Falco tinnunculus*) in south Scotland. *J. Zool.*, 206, 175–89.

Village, A. (1990). *The kestrel*. London: T. and A. D. Poyser.

Village, A., Marquiss, M. and Cook, D. C. (1980). Moult, ageing and sexing of kestrels. *Ring. Migrat.*, 3, 53–9.

Villarroel, M., Bird, D. M. and Kuhnlein, U. (1998). Copulatory behaviour and paternity in the American kestrel: the adaptive significance of frequent copulations. *Anim. Behav.*, 58, 289–99.

Vincent, J. (1966). *Red Data Book – Aves*. Morges: International Union for Conservation of Nature and Natural Resources.

Visser, M. E., Holleman, L. J. M. and Caro, S. P. (2009). Temperature has a causal effect on avian timing of reproduction. *Proc. R. Soc. Lond. B*, 276, 2323–31.

von Schantz, T., Bensch, S., Grahn, M., Hasselquist, D. and Wittzell, H. (1999). Good genes, oxidative stress and condition-dependent sexual signals. *Proc. R. Soc. Lond. B*, 266, 1–12.

Wagner, N. D., Hillebrand, H., Wacker, A. and Frost, P. C. (2013). Nutritional indicators and their uses in ecology. *Ecol. Lett.*, 16, 535–44.

Walker, L. A., Shore, R. F., Turk, A., Pereira, M. G. and Best, J. (2008). The Predatory Bird Monitoring Scheme: identifying chemical risks to top predators in Britain. *Ambio*, 37, 466–71.

Wallin, K. (1984). Decrease and recovery patterns of some raptors in relation to the introduction and ban of alkyl-mercury and DDT in Sweden. *Ambio*, 13, 263–5.

Wallin, K., Järås, T., Levin, M., Strandvik, P. and Wallin, M. (1983). Reduced adult survival and increased reproduction in Swedish kestrels. *Oecologia*, 60, 302–5.

Wallin, K., Wallin, M. L., Järås, T. and Strandvik, P. (1987). Leap-frog migration in the Swedish kestrel *Falco tinnunculus* population. In M. O. G. Eriksson, ed., *Proceedings of the fifth Nordic ornithological congress* (pp. 213–22). Goteborg: Kungl. Vetenskaps- och Vitterhets-Samhallet.

Wang, Y., Liu, H., Wang, H., Ma, L. and Yi, G. (2019). Polygyny in the Eurasian kestrel (*Falco tinnunculus*): behavior, morphology, age, heterozygosity, and relatedness. *J. Rapt. Res.*, 53, 202–6.

Ward, F. P., Fairchild, D. G. and Vuicich, J. V. (1971). Inclusion body hepatitis in a prairie falcon. *J. Wildl. Dis.*, 7, 120–4.

Watson, J. (1992). Nesting ecology of the Seychelles Kestrel *Falco araea* on Mahé, Seychelles. *Ibis*, 134, 259–67.

Watson, J. (1981). *Population ecology, food and conservation of the Seychelles kestrel Falco araea on Mahé*. PhD thesis, University of Aberdeen, Aberdeen, Scotland.

Watson, J. (1993). Breeding cycle of the Seychelles Kestrel. In M. K. Nicholls and R. Clarke, eds., *Biology and conservation of small falcons* (pp. 73–9). London: Hawk and Owl Trust.

White, C., Olsen, P. and Kiff, L. (1994). Family Falconidae (Falcons and Caracaras). In J. del Hoyo, A. Elliott and J. Sargatal, eds., *Handbook of the birds of the world*. Vol. II (pp. 216–75). Barcelona: Lynx Edicions.

Whitney, M. C. and Cristol, D. A. (2018). Impacts of sublethal mercury exposure on birds: a detailed review. *Rev. Environ. Contam. Toxicol.*, 244, 113–63.

Whittingham, M. J. and Devereux, C. L. (2008). Changing grass height alters foraging site selection by wintering farmland birds. *Basic Appl. Ecol.*, 9, 779–88.

Wiebe, K. (1996). The insurance-egg hypothesis and extra reproductive value of last-laid eggs in clutches of American kestrels. *The Auk*, 113, 258–61.

Wiebe, K. L. and Bortolotti, G. R. (1992). Facultative sex ratio manipulations in American kestrels. *Behav. Ecol. Sociobiol.*, 30, 379–86.

Wiebe, K. L. and Bortolotti, G. R. (1993). Brood patches of American kestrels: an ecological and evolutionary perspective. *Ornis Scand.*, 24, 197–204.

Wiebe, K. L. and Bortolotti, G. R. (1994). Energetic efficiency of reproduction: the benefits of asynchronous hatching for American kestrels. *J. Anim. Ecol.*, 63, 551–60.

Wiebe, K. L. and Bortolotti, G. R. (1995). Egg size and clutch size in the reproductive investment of American kestrels. *J. Zool.*, 237, 285–301.

Wiebe, K. L. and Bortolotti, G. R. (1996). The proximate effects of food supply on intraclutch egg-size variation in American kestrels. *Can. J. Zool.*, 74, 118–24.

Wiebe, K. L., Korpimäki, E. and Wiehn, J. (1998a). Hatching asynchrony in Eurasian kestrels in relation to the abundance and predictability of cyclic prey. *J. Anim. Ecol.*, 67, 908–17.

Wiebe, K. L., Wiehn, J. and Korpimäki, E. (1998b). The onset of incubation in birds: can females control hatching patterns? *Anim. Behav.*, 55, 1043–52.

Wiebe, K. L., Jönsson, K. I., Wiehn, J., Hakkarainen, H. and Korpimäki, E. (2000). Behaviour of female Eurasian kestrels during laying: are there time constraints on incubation? *Ornis Fenn.*, 77, 1–9.

Wiehn, J. (1997). Plumage characteristics as an indicator of male parental quality in the American Kestrel. *J. Avian Biol.*, 28, 47–55.

Wiehn, J. and Korpimäki, E. (1997). Food limitation on brood size: experimental evidence in the Eurasian kestrel. *Ecology*, 78, 2043–50.

Wiehn, J. and Korpimäki, E. (1998). Resource levels, reproduction and resistance to haematozoan infections. *Proc. R. Soc. Lond. B*, 265, 1197–201.

Wiehn, J., Korpimäki, E. and Pen, I. (1999). Haematozoan infections in the Eurasian kestrel: effects of fluctuating food supply and experimental manipulation of parental effort. *Oikos*, 84, 87–98.

Wiehn, J., Ilmonen, P., Korpimäki, E., Haataja, M. and Wiebe, K. (2000). Hatching asynchrony in the Eurasian kestrel *Falco tinnunculus*: an experimental test of the brood reduction hypothesis. *J. Anim. Ecol.*, 69, 85–95.

Wiemeyer, S. N. and Porter, R. D. (1970). DDE thins eggshells of captive American kestrels. *Nature*, 227, 737–8.

Wiklund, C. G. and Village, A. (1992). Sexual and seasonal variation in territorial behaviour of kestrels, *Falco tinnunculus*. *Anim. Behav.*, 43, 823–30.

Williams, M., Krootjes, B. B., van Steveninck, J. and van der Zee, J. (1994). The pro- and antioxidant properties of protoporphyrin IX. *Biochim. Biophys. Acta*, 1211, 310–6.

Williams, T. D. (1994). Intraspecific variation in egg size and egg composition in birds: effects on offspring fitness. *Biol. Rev.*, 68, 35–59.

Willoughby, E. J. and Cade, T. J. (1964). Breeding behaviour of the American kestrel (sparrow hawk). *Living Bird*, 3, 75–96.

Wilson, R. P., White, C. R., Quintana, F., et al. (2006). Moving towards acceleration for estimates of activity-specific metabolic rate in free-living animals: the case of the cormorant. *J. Anim. Ecol.*, 75, 1081–90.

Wingfield, J. C., Maney, D. L., Breuner, C. W., et al. (1998). Ecological bases of hormone-behavior interactions: the 'emergency life history stage'. *Am. Zool.*, 38, 191–206.

Wink, M. and Sauer-Gürth, H. (2004). Phylogenetic relationships in diurnal raptors based on nucleotide sequences of mitochondrial and nuclear marker genes. In R. D. Chancellor and B.-U. Meyburg, eds., *World working group on birds of prey* (pp. 483–98). Budapest: Berlin and MME/BirdLife Hungary.

Work, T. H., Hurlbut, H. S. and Taylor, R. M. (1955). Indigenous wild birds of the Nile Delta as potential West Nile circulating reservoirs. *Am. J. Trop. Med. Hyg.*, 4, 872–8.

Yalden, D. W. and Yalden, P. E. (1985). An experimental investigation of examining Kestrel diet by pellet analysis. *Bird Study*, 32, 50–5.

Yilmaz, S. and Yilmaz, E. (2006). Effects of melatonin and vitamin E on oxidative–antioxidative status in rats exposed to irradiation. *Toxicol.*, 222, 1–7.

Yoda, K., Sato, K., Niizuma, Y., et al. (1999). Precise monitoring of porpoising behaviour of Adelie penguins determined using acceleration data loggers. *J. Exp. Biol.*, 202, 3121–6.

Young, G. R., Dawson, A., Newton, I. and Walker, L. (2009). The timing of gonadal development and moult in three raptors with different breeding seasons: effects of gender, age and body condition. *Ibis*, 151, 654–66.

Zahavi, A. (1975). Mate selection: a selection for a handicap. *J. Theor. Biol.*, 53, 205–14.

Zahavi, A. and Zahavi, A. (1997). *The handicap principle: a missing piece of Darwin's puzzle*. Oxford, UK: Oxford University Press.

Zampiga, E., Gaibani, G. and Csermely, D. (2008a). Sexual dimorphism colours and female choice in the common kestrel. *Ital. J. Zool.*, 75, 155–9.

Zampiga, E., Gaibani, G. and Csermely, D. (2008b). Ultraviolet reflectance and female mating preferences in the common kestrel (*Falco tinnunculus*). *Can. J. Zool.*, 86, 479–83.

Zeman, M., Výboh, P., Juráni, M., et al. (1993). Effects of exogenous melatonin on some endocrine, behavioural and metabolic parameters in Japanese quail *Coturnix coturnix japonica*. *Comp. Biochem. Physiol. Part A*, 105, 323–8.

Zhang, L., Liu, Y. and Song, J. (2008). Genetic variation between subspecies of Common Kestrels (*Falco tinnunculus*) in Beijing, China. *J. Raptor Res.*, 42, 214–9.

Zink, R. (1998). Fortpflanzungsstrategien kolonialer und solitärer Turmfalken (*Falco tinnunculus*). Diplomarbeit, Universität Wien.

Żmihorski, M. and Rejt, Ł. (2007). Weather-dependent variation in the cold-season diet of urban kestrels *Falco tinnunculus*. *Acta Ornithol.*, 42, 107–13.

Zombor, K. and Tóth, M. (2015). Mivel táplálkozik a vörös vércse (*Falco tinnunculus* Linnaeus, 1758) Budapesten? *Állattani Közlemények*, 100, 111–34.

Zuberogoitia, I., Zabala, J. and Martínez, J. E. (2018). Moult in birds of prey: a review of current knowledge and future challenges for research. *Ardeola*, 65, 183–207.

Zuk, M., Ligon, J. D. and Thornhill, R. (1992). Effects of experimental manipulations of male secondary sex characters on mate preference in red jungle fowl. *Anim. Behav.*, 44, 999–1006.

Index